Plant and Animal Endemism in California

Susan P. Harrison

UNIVERSITY OF CALIFORNIA PRESS
Berkeley · Los Angeles · London

University of California Press, one of the most
distinguished university presses in the United States,
enriches lives around the world by advancing scholarship
in the humanities, social sciences, and natural sciences.
Its activities are supported by the UC Press Foundation
and by philanthropic contributions from individuals and
institutions. For more information, visit www.ucpress.edu.

University of California Press
Berkeley and Los Angeles, California

University of California Press, Ltd.
London, England

© 2013 by The Regents of the University of California

Library of Congress Cataloging-in-Publication Data

Harrison, Susan (Susan Patricia)
    Plant and animal endemism in California / Susan Harrison.
      pages cm
    Includes bibliographical references and index.
    ISBN 978-0-520-27554-6 (cloth : alkaline paper)
    1. Endemic plants—California.   2. Endemic
animals—California.   3. Endemic plants—Ecology—
California.   4. Endemic animals—Ecology—California.
5. Endemic plants—Conservation—California.
6. Endemic animals—Conservation—California.   I. Title.
  QK149.H444 2013
  581.9794—dc23
                                   2012043370

Manufactured in the United States of America

22 21 20 19 18 17 16 15 14 13

10 9 8 7 6 5 4 3 2 1

The paper used in this publication meets the minimum
requirements of ANSI/NISO Z39.48–1992 (R 2002)
(*Permanence of Paper*).

Cover image: Sickle-leaved onion (*Allium falcifolium*),
Oregon stonecrop (*Sedum oreganum*), and Siskiyou
bitterroot (*Lewisia cotyledon*), members of three genera
rich in Californian endemics.

Contents

Preface and Acknowledgments

Why write about endemism in California? Asked by a group of friendly but critical graduate students during a seminar visit, this question struck me as worth thinking about before beginning to write. Ecologists and evolutionists usually focus on broad, theory-driven questions. Isn't it small-minded to focus on a place, let alone a place defined by human-drawn boundaries? And what about endemism, which the dictionary defines as "the condition of being native or restricted to a certain place"? Every species is endemic to somewhere, so what makes Californian endemism interesting?

My answers will be familiar to many readers, but perhaps not to all, as my experience with the out-of-state graduate students suggests. California is not just a political unit; uniquely among U.S. states, it is also more or less its own biogeographic region. More precisely, the state largely coincides with the California Floristic Province, one of only five regions in the world where the mediterranean climate is found. (This book considers endemism in the California Floristic Province wherever possible but concentrates largely on endemism in the state of California simply because of the greater availability of data at the state level.) The mediterranean biome worldwide is outstanding for its botanical uniqueness; it holds an estimated 20 percent of the world's vascular plants in only 2 percent of the world's land area. Some of this biome's most distinctive groups of plants are thought to be evolving rapidly, and many classic studies of plant evolution and speciation have emerged from

California. Finally, California has been the site of scientific and policy experiments aimed at the conservation of biological diversity, in part because traditional approaches to conservation are challenged by the sheer abundance and diffuse distribution of rare species in the state.

This book is motivated by the aim to learn new lessons at the interface of evolution, ecology, and conservation by examining California. Thus it focuses on analyzing patterns and addressing general questions, as outlined in the introduction. What this book does not do is explore the state's rich natural history in any great depth; that has already been done well by many authors. Readers are directed, for example, to Elna Bakker's *An Island Called California,* Allan Schoenherr's *Natural History of California* and Schoenherr and colleagues' *Natural History of the Islands of California,* Peter Dallman's *Plant Life in the World's Mediterranean Climates,* and the entire California Natural History Guides series published by UC Press. Key resources for data on the state's species and habitats are the Department of Fish and Wildlife (www.dfg.ca.gov/biogeodata), the Native Plant Society (www.rare-plants.cnps.org), and the Calflora project (www.calflora.org).

The botanical bias of this book has to be admitted at the outset. Reasons for this are probably obvious: the long history of studying endemism in California and even of treating California as a biogeographic unit has been largely the work of plant-oriented scientists. Not coincidentally, as Chapters 3 and 4 discuss, Californian endemism is more pronounced in plants than in most animal groups. It is no coincidence, then, that a plant person should attempt a book on Californian endemism. Your author freely admits to having studied Californian plant diversity for the past fifteen years, although my degrees are in zoology, ecology, and biology, and my master's and PhD work was on insect ecology. Please don't close the cover, animal lovers; you will find here never before compiled material on the state's winged, finned, and four- to eight-legged inhabitants. Contemplating the contrasts between plant and animal endemism has been an enjoyable exercise that I hope will interest you as well.

This book was made possible by the generosity of many people. Expert knowledge and data came from Bruce Baldwin, Roxanne Bittman, David Bunn, Frank Davis, Tom Gardali, Terry Griswold, Brenda Johnson, Doug Kelt, Lynn Kimsey, Tim Manolis, Richard Moe, Peter Moyle, Paul Opler, Monica Parisi, Jerry Powell, Greg Pauly, Gordon Pratt, Jim Quinn, Steve Schoenig, Nathan Seavy, Art Shapiro, Aaron Sims, Robert Thomson, James Thorne, Robbin Thorp, Darrell

Ubick, Dirk Van Vuren, Phil Ward, David Wake, and David Weissman. The new list of California Floristic Province endemic plants was generously created by Dylan Burge, and the Baja California data were kindly updated by Bart O'Brien. Artwork (or data for artwork) was generously provided by Bruce Baldwin, Ron Blakey, Dylan Burge, Paul Fine, Kathy Keatley Garvey, Brad Hawkins, Lynn Kutner, Ryan O'Dell, Brody Sandel, Cristina Sandoval, Aaron Schusteff, Zack Steel, David Wake, David Weissman, Darrell Ubick, Joseph Vondracek, and James Zachos. Assistance with data processing was given by Brian Anacker, Erica Case, and Brandon Sepp, and illustration help was provided by Steven Oerding and Marko Spasojevic. Kind yet helpful comments on early drafts were given by Howard Cornell, Peter Moyle, Philip Rundel, Mark Stromberg, and John Thompson. Chuck Crumly and Lynn Meinhardt at UC Press shepherded this book through its many stages. Finally, this book is dedicated to all those who have worked to understand and conserve the "Californian" (in the broad sense) flora and fauna . . . you know who you are!

Introduction

Endemism or biological uniqueness is woven into what most people think of when they hear "California." Along with Hollywood, the Golden Gate Bridge, and wineries, even nonbiologists might think of coastal redwoods, condors, and fields of orange poppies in their imaginings of the state. Almost anything famously Californian has some connection to the wealth of unique species. The pleasantly winter-wet/summer-dry (mediterranean) climate, for example, is found in only five places in the world; it invariably means not only excellent wines and dense human populations, but an abundance of native plant species adapted to the long dry season. Hollywood is named after an endemic plant with the characteristic mediterranean climate trait of thick ever-green leaves (toyon, *Heteromeles californica,* one of over 300 plants with the species name *californica* or *californicus*). The fog shrouding the coastal redwoods hints at an ancient, wetter, less seasonal climate that also figures importantly in explaining the state's biological riches. Then there's geology; an active plate tectonic margin laid the ground-work, literally, for California's frequent earthquakes and history-making gold rush and the granite pinnacles of Yosemite. Geologic forces also created the array of past and present barriers, including the Golden Gate, Monterey Bay, mountain ranges, and deserts, that not only gave the state spectacular scenery but also split ancestral plant and animal lineages into today's diverse suites of species. Geologic upheaval also gave rise to the state's dramatic variety in climates, bedrock, and soils, producing the most diverse agricultural region in the world and creating equally diverse habitats for native species.

Some characteristically Californian strains of intertwined nature and culture are found in the state's problems and conflicts, as well as its attractions. Invasive non-native species, threatened and endangered species, and rare plants with disputed taxonomic status are more numerous in California than in most other world regions. Wildfires have become more frequent and severe in Southern California in recent decades, intensifying the challenges of conserving natural habitats in urbanizing landscapes. Water wars, a perennial feature of California politics, have intensified as endemic fish have crashed from fantastic abundance to near-extinction and have been officially listed as endangered. Climate change has led to proposals for alternative energy, mass public transportation, and water storage that threaten some of the state's best remaining natural areas and longest-held conservation priorities. None of these problems is unique to California, of course. However, their sheer intensity stems from the state's rich biological diversity combined with its dense human population and, in turn, from its particular blend of climate, topography, and geology.

Focusing on endemism is not the only approach to understanding the origins, distribution, or conservation of biological diversity, but endemism offers an alluring pathway into these questions. Evolutionary biologists have often studied newly evolved endemics to understand the origin of species, and biogeographers use ancient relict endemics as clues to past environments. For ecologists such as me, a biologically distinct region such as California offers the chance to explore how large-scale evolutionary and historical forces (climate change, plate tectonics) interact with local ecological processes (dispersal, competition, disturbance) to assemble ecological communities. For conservation biologists, analyses of geographic concentrations, or hotspots, of endemism have been an important aspect of designing effective strategies. In the policy realm, no law has been more important for biological conservation than the U.S. Endangered Species Act (ESA), and the sheer abundance of ESA-listed species in California (around half of them restricted to the state) has motivated many attempts to fine-tune and supplement this cornerstone law.

Why are there so many endemic species in California? From a scientific perspective, the problem is not so much to find a plausible explanation for California's richness of endemics as to choose among too many compelling explanations. One often reads that California's biological

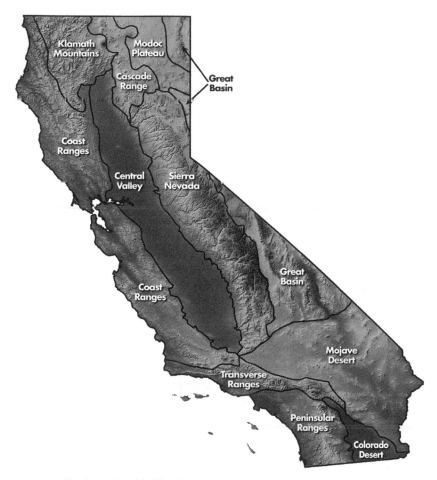

FIGURE 1. Physiography of California.

richness is a result of its tremendous heterogeneity, in other words, its broad spans of elevation, latitude, and coastal-to-interior climates, the soil variation caused by its complex geologic structure, and the resulting rich diversity of vegetation types. Though environmental variability is certainly one explanation for high endemism, there are others. California is also rich in internal barriers to dispersal, including mountain ranges, waterways, and offshore islands that have appeared and (in some cases) disappeared over the past 50 million years, leaving detectable imprints on today's species and genetic diversity (Figure 1).

The mediterranean climate seems almost indisputably linked with California's botanical richness. In this odd climate, as in no other, the two

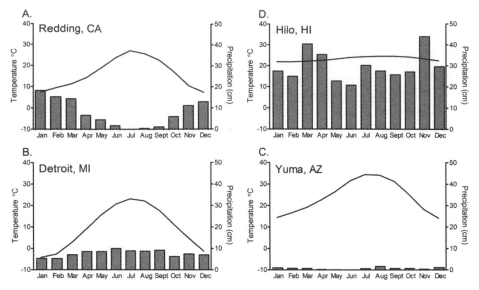

FIGURE 2. Seasonal distribution of rainfall and temperature in mediterranean climate (Redding, CA); north-temperate climate (Detroit, MI); desert climate (Yuma, AZ); and tropical climate (Hilo, HI).

things that plants need most—rainfall and warm growing temperatures—are almost completely decoupled from one another in the course of the year (Figure 2). A few fleeting weeks of ideal growing conditions in spring are bracketed by cool, rainy winters and fiercely long, dry summers. Plants adapt in varied ways. Many herbs grow slowly or not at all in winter, mature rapidly and flower in spring, and survive summer as dormant seeds, bulbs, or roots. Lacking these options, trees and shrubs endure summer drought by having tough evergreen leaves or by shedding leaves in the summer. Hot, dry summers followed by windy falls generate intense fires, and plants either resprout or regenerate from dormant seeds. These strategies have evolved in the floras of all five of the world's mediterranean climate regions, all of which are rich in endemic plants (Figure 3). But climatic history may be as important as today's climate in explaining California's biotic uniqueness. The region has remained somewhat equable throughout the global cooling and drying of the past 50 million years (see Chapter 2), avoiding the extremes of glaciation and desertification that have affected much of the earth's terrestrial surface.

This book aims to examine all these factors—environmental heterogeneity, barriers, contemporary climate, and climate history—as

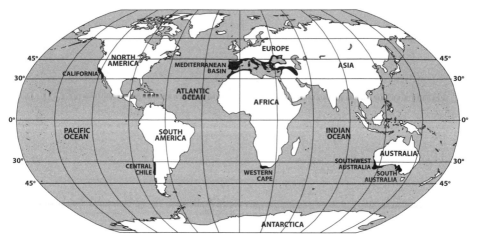

FIGURE 3. World distribution of mediterranean climates.

explanations for California's endemic richness. Instead of being satis-
fied with the conclusion that they are all important, an attempt is made
to evaluate them critically, using many sources of evidence: comparisons of California with other parts of North America, comparisons of
the five mediterranean climate regions to one another and the rest of the
world, comparisons of species richness and endemism in different
regions within California, and evidence from evolutionary studies of
Californian plants and animals.

The stage is set in Chapter 1 by considering the meaning of endemism, the finer points and pitfalls of measuring endemism, broad global
patterns of species diversity and endemism, and the general modes by
which species become endemics. The physical history of California and
the classic story of the origins of its endemic-rich flora are reviewed in
Chapter 2. In Chapter 3, the questions are posed, for plants, Does the
classic story hold up under new evidence? What are the relative roles
of physical heterogeneity, the novel mediterranean climate, internal
barriers, and long-term climatic stability in producing plant endemism
in California? Animals are the subject of Chapter 4, which asks what
levels of endemism are found in various animal groups in California
and whether the explanations relevant to plants also hold up for
animals. Chapter 5 examines the unique challenges of conservation in
an endemic-rich region and how these are being met in California. The
book closes with an attempt to synthesize the answers (Chapter 6).

1

Biotic Uniqueness

An Overview

Endemism, or the confinement of species or other taxa to particular geo-graphic areas, can be a slippery concept. Every species is confined to some place; for example, it has been estimated that more than 90 percent of the world's plant species are found in only one floristic province (Kruckeberg and Rabinowitz 1985). So when do species or places become interesting on account of their "endemism"? Islands with unique floras and faunas provide the clearest answer. It is no accident that the Galápagos were instrumental to Darwin's thinking. Long-distance colo-nization, the curtailment of gene flow with close relatives, adaptation to new biotic and abiotic conditions, and (in some cases) the survival of ancestral forms that have become extinct on mainlands can be seen and studied with exceptional clarity on islands that are rich in species found nowhere else. Similar evolutionary forces may be revealed to operate more subtly in regions and habitats with islandlike qualities. California is a good example of an islandlike area within a continent; it is a region of mediterranean climate completely surrounded by mountains, desert, and ocean hostile to much of its flora and fauna, and the nearest similar "islands" are far away, in Chile and the Mediterranean Basin.

The endemic-rich Californian flora has been an influential living laboratory for the study of plant adaptation and speciation. Two of the founders of modern plant evolutionary biology were G. Ledyard Stebbins (1906–2000; UC Berkeley and UC Davis), who first focused evolutionary theory on the study of plants with his *Variation and*

Evolution in Plants (1950) and whose work called attention to the central roles of hybridization and polyploidy in plant speciation; and Jens Clausen (1891–1969; Carnegie Institution), who is best known for leading interdisciplinary experimental studies of genetic differentiation of plant populations along gradients and who wrote *Stages in the Evolution of Plant Species* (1951). Since the mid-twentieth century, there has been a flourishing tradition of using endemic-rich Californian genera such as *Clarkia, Ceanothus, Limnanthes, Madia,* and *Mimulus* as model systems in evolutionary biology (see Chapter 3).

PROBLEMS IN DEFINING ENDEMISM

Before discussing endemism, or geographic restriction, of species to either the state of California or the California Floristic Province (CFP), let us consider some of the issues that affect its definition.

Relationship to Rarity

In common with many other works, this book uses the term *endemism* to mean the condition of having a limited geographic range, regardless of whether a species can be considered rare. However, in the literature on the biology of rarity, the term is sometimes used in a narrower sense. For example, in a classic review of endemism in higher plants, Kruckeberg and Rabinowitz (1985) define *endemics* as species existing as only one or a few populations. They note that such species can nearly always be considered rare in the sense of having very small geographic ranges. Many endemics (as defined by these authors) are also rare in the sense of having narrow niches; the best-known examples are plants specialized on particular soils, often called "edaphic endemics." Endemism is uncorrelated with a third type of rarity, namely, low population density; these authors note that endemics are often locally abundant within their narrow geographic ranges or habitats.

Appropriate Spatial Units

Islands are natural units for defining and measuring endemism, because the boundaries of an island are clearly defined and obviously linked to the evolutionary processes giving rise to unique species. This is less true for almost any other kind of geographic unit. Political boundaries seem especially inappropriate since they are unrelated to biology, yet the majority of the world's biodiversity data are compiled by country, state, province, or other similar unit. In the United States, an important source of data is the Natural Heritage Network, a national program founded by the Nature Conservancy in the mid-1970s and now implemented by each state. Each member of the network—in California's case, its Department of Fish and Wildlife—compiles occurrence records of imperiled species and other conservation elements such as natural communities and makes these records available in an interchangeable format. Analyses of these data (Stein et al. 2000), discussed in Chapters 3 and 4, point to California as the U.S. state with the highest number of total and endemic species, although Hawaii is higher in percentage endemism, as is often true of oceanic islands. The problem with this state-based approach is that it greatly understates the diversity of biogeographic regions that occur across many states. Appalachia is an important U.S. center of biodiversity and endemism that encompasses eight states, none of which ranks particularly high in state-level analyses.

Ecoregions are units defined by biogeographers on the basis of shared climates, vegetation types, and major assemblages of species. Various classifications are used by conservationists (e.g., World Wildlife Fund, Nature Conservancy), resource managers (e.g., U.S. Forest Service, Environmental Protection Agency), and biological databases (e.g., *The Jepson Manual* [Baldwin et al. 2012]). Analyzing endemism by ecoregions seems more defensible than by states, but it has its pitfalls too, and California is a good example. In a global conservation assessment (Ricketts et al. 1999), California does not register high in either species diversity or endemism; as the authors acknowledge, this is because California is so diverse that it is divided into 13 ecoregions. If California's biological uniqueness results from a heterogeneous landscape, across which a common ancestral pool of species has diverged into many localized endemics, this approach underestimates the "true" diversity of California to the same extent that the state-based approach underestimates Appalachia.

Biogeographic units based on assemblages of related species are another alternative. In the most widely used system, the California Floristic Province forms part of the Madrean Region, which belongs in turn to the Holarctic Kingdom (Table 1; Takhtajan 1986). The majority of authors define the California Floristic Province as including all nondesert parts of the state of California, plus south-central Oregon and northwestern Baja California (Figure 4; see, e.g., Raven and Axelrod 1978; Conservation International 2011; Baldwin et al. 2012).

TABLE I   FLORISTIC KINGDOMS, REGIONS, AND PROVINCES

Holarctic Kingdom
    Madrean Region
    Californian Province
    Sonoran, Mexican Highlands, and Great Basin Provinces
    Circumboreal, Eastern Asiatic, North American Atlantic, Rocky Mountain,
       Macaronesian, Mediterranean, Saharo-Arabian, and Irano-Turanian Regions
Paleotropical Kingdom
    Guineo-Congolian, Usambara-Zululand, Sudano-Zambezian, Karoo-Namib,
       St. Helena and Ascension, Madagascan, Indian, Indochinese, Malesian, Fijian,
       Polynesian, Hawaiian, and Neocaledonian Regions
Neotropical Kingdom
    Caribbean, Guayana Highlands, Amazonian, Brazilian, and Andean Regions
South African Kingdom
    Cape Region
Australian Kingdom
    Northeast Australian, Southwest Australian, and Central Australian Regions

SOURCE: From Takhtajan 1986.

Under a narrower definition, the wetter areas of northwestern California and southern Oregon may be considered part of the Rocky Mountain Province (Takhtajan 1986). The California Floristic Province broadly coincides with the mediterranean climate or mediterranean biome, as defined by rainy winters, dry summers, annual precipitation of 25 to 100 centimeters, and sclerophyllous vegetation (Dallman 1998). (Again, by some definitions, northwestern California and southern Oregon are too rainy, the Sierra Nevada too snowy, and parts of the Central Valley too arid to be considered mediterranean.) Under the broad definition, which is consistent with a floristic analysis of the West Coast (Peinado et al. 2009), there is 70 percent geographic overlap between the state of California and the California Floristic Province (Conservation International 2011). Thus it is reasonable to speak of endemism in California as a natural phenomenon and not just the product of a political boundary. This book uses the broad definition of the California Floristic Province (see Figure 4), in accordance with major works on the flora (Raven and Axelrod 1978; Baldwin et al. 2012), and a novel effort is made to compile data on its animal endemism.

Data on endemism in the state of California were generally obtained from published sources (plants, Baldwin et al. 2012; mammals, CDFW 2003; birds, Shuford and Gardali 2008; reptiles and amphibians,

FIGURE 4. The California Floristic Province, under a broad definition that includes the Sierra Nevada, the northern Coast Ranges, and parts of the Central Valley (regions sometimes excluded from the CFP as being too cold, wet, or dry to be "mediterranean").

Jennings and Hayes 1994; fish, Moyle 2002; butterflies, Pelham 2008); these lists were updated for taxonomic and distributional changes by consultation with experts. Data on endemism in the California Floristic Province were harder to obtain. Remarkably, for plants there is currently no database from which the thousands of species endemic to the Floristic Province can easily be counted or identified, but a preliminary attempt is made in this book (see Appendix). For animals, the modest lists of Floristic Province endemics were obtained by visually interpreting range maps in atlases and by asking experts on each group.

Spatial and Taxonomic Scales

Systematic biases in the estimation of endemism arise from both spatial and taxonomic scales. Larger geographic units will tend to have more endemics than smaller ones. Converting numbers of species to species density (species/area), as is sometimes done, is not a valid correction for this bias because the expected number of species (S) does not increase linearly with area (A). Instead, it follows a logarithmic relationship, $S = cA^z$, where the exponent z is typically 0.15–0.35 among islands or other units that share some of their species. A tenfold increase in area therefore results in only an approximate doubling of species, and species density (S/A) has a strong bias toward being higher on small islands. Among continents or other units sharing relatively few species, z may approach 1.0, reducing the bias in species density (Rosenzweig 1995). Still, the best way to correct diversity for variation in area is to use S/A^z, where z is estimated from regressing $\ln(S)$ on $\ln(A)$. Another solution is to calculate diversity and endemism from species range maps that have been converted to equal-area polygons (e.g., Stein et al. 2000; CDFW 2003), as long as the underlying data are accurate enough.

With regard to taxonomic scale, some data sources report endemism based on all named taxa (species, subspecies, and varieties); others report only full species. Logically, endemism in a given geographic area will always be higher among taxa of lower rank (Kruckeberg and Rabinowitz 1985). Taxa below the species level are described more often and on the basis of smaller differences in vertebrates than invertebrates, and in showier invertebrates (butterflies) than inconspicuous ones (most others). Examples from California suggest this leads to considerable bias. In kangaroo rats, 23 subspecies but only 5 full species are endemic to the state (Goldingay et al. 1997). In birds, 64 named taxa but only 2 full species are state endemics (Shuford and Gardali 2008). In plants, however, endemism is 34 percent for all named taxa and 28 percent for full species (Chapter 3, Table 3). Grasshoppers show endemism of 53 percent for full species plus subspecies and 51 percent for full species only (Chapter 4). The much smaller disparities for plants and grasshoppers than for kangaroo rats and birds suggests that subspecies and varieties are less often described in plants and invertebrates than in vertebrates. In the majority of invertebrates, in fact, surveys are too incomplete for even crude estimates of species-level endemism (Chapter 4). Full species are the focus of this book because of the extra subjectivity and bias introduced by subspecies.

Defining species remains a perennial source of debate in both plant and animal systematics (Mallet 2001). Traditionally, most taxonomists have sought consistent breakpoints in the variation of multiple traits, presumably reflecting a lack of gene flow, as a way to define the boundaries between related species (e.g., Oliver and Shapiro 2007). As molecular data have become increasingly available, one alternative that has gained popularity is that any unique trait can define a lineage as a species (Mallet 2001). In practice, these diagnostic traits are often variations in mitochondrial DNA, which evolves relatively fast in animals. Many existing species can be split up into multiple, small-ranged, and morphologically nearly identical new species under this concept (Agapow et al. 2004). The California raven, for example, could be its own species based on molecular variation, even though it does not differ in appearance or behavior from other North American ravens (Omland et al. 2000). Species numbers would more than double in plants and nearly double in most groups of animals under this "phylogenetic" or "diagnostic" species concept (Agapow et al. 2004), leading to even more substantial increases in endemism. This book accepts and includes all species that have been formally described by any method but does not deal with proposed new species of unclear status, nearly all of which are subdivisions of existing species.

Relative versus Absolute Values

Endemism may be reasonably expressed and compared either in percentages or numbers of species. It is worth remembering that percentages are more meaningful the greater the diversity as well as the higher the taxonomic rank of the group being examined. Thus 50 percent Californian endemism in the grasshopper family Acrididae (with 186 species in the state) means more for the state's biotic uniqueness than 50 percent endemism in the grasshopper families Eumastacidae and Tanaoceridae (4 and 2 species in the state), or even than 86 percent endemism in the 21 species of *Timema* (a genus of walking stick insects). Throughout this book, endemism is expressed in both numbers and percentages, in the belief that they provide complementary information.

Comparative information from other geographic regions is essential to characterizing and explaining Californian endemism. Acridid grasshoppers are one of the most endemic-rich groups in California, but they may be equally so in other parts of the mountainous western United States (Knowles and Otte 2000). Whether or not it is remarkable that

5 of 23 kangaroo rats (*Dipodomys*) or 21 of 22 slender salamanders (*Batrachoseps*) are endemic to California depends on whether ecologically similar groups are just as diverse in neighboring regions. It is challenging to find, for almost any group, either comparative data or interpretive analyses that place endemism in California in a larger context. This book relies on comparisons with other states and the other four mediterranean climate regions to provide a context for Californian endemism.

LARGE-SCALE PATTERNS IN SPECIES RICHNESS AND ENDEMISM

One of the best predictors of species richness at a global scale is plant productivity, which is determined at large scales by the abundance of water and solar energy. At low latitudes water exerts stronger control, whereas at high latitudes solar energy is a stronger limitation. There are consistently more species of plants and animals in the warm and wet parts of the world than the colder or drier ones, regardless of whether the latitude is tropical or nontropical (Figure 5a; Hawkins et al. 2003). Within the United States as a whole, plant and vertebrate animal diversity is higher in the warmer southerly states (Stein et al. 2000). Within California, in contrast, the diversities of plants, birds, mammals, and amphibians (although not reptiles) are highest in the rainier north (CDFW 2003). However, this is a case where the exception proves the rule, because California is a sunny but arid region in which water is the limiting factor governing plant productivity. Plant diversity in California is positively related to a remotely sensed index of productivity, which in turn is strongly related to rainfall but not to temperature (Figure 5b; Harrison et al. 2006).

Levels of endemism may follow geographic patterns different from total species diversity. Isolated islands, for example, are often high in endemism but low in total richness. Endemism on continents is harder to explain, but one recent analysis suggests that global patterns in endemism are best explained by climatic stability. Sandel et al. (2011) defined the climate change velocity of any given location as the ratio of climatic change over time at that location since the last glacial maximum, 22,000 years ago, to the average change in climate over space at the same location at present. This velocity represents how fast an organism has had to shift its distribution to keep pace with postglacial warming. It is slow, for example, in mild maritime climates that have undergone less change in temperature over time and in rugged regions where present-day

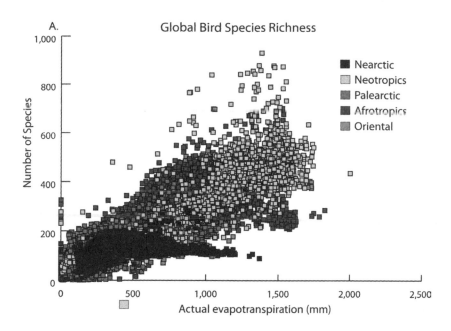

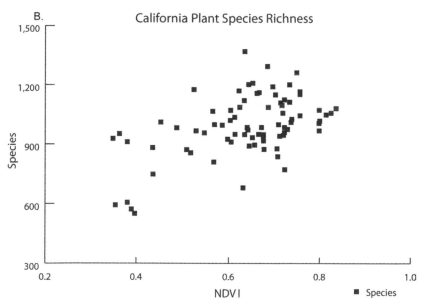

FIGURE 5. Large-scale species diversity as a function of two different measures of plant productivity.

(a) Global relationship between bird species richness and actual evapotranspiration (AET), an index that combines growing season temperature and moisture (data from B.A. Hawkins unpublished; see also Hawkins and Porter 2003).

(b) Californian plant species richness versus the normalized difference vegetation index (NDVI), a remotely sensed metric of "greenness," or the red/far red reflectance ratio that correlates strongly with mean annual rainfall in California (data from Harrison et al. 2006).

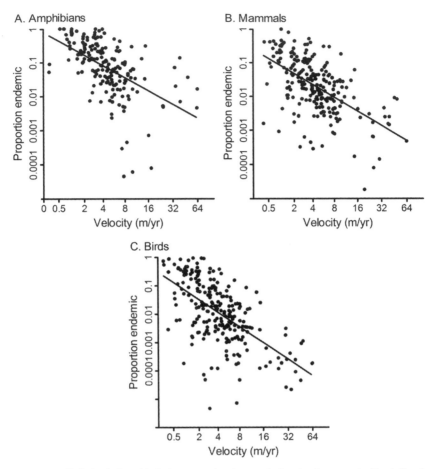

FIGURE 6. Global relationship between endemism and climate change velocity, defined as the rate at which species must migrate to remain in a constant temperature since the last glacial maximum. Climate change velocity is calculated by dividing the rate of temperature change through time by the local rate of change over space. (Data from B. Sandel unpublished)

temperatures vary sharply over short distances (e.g., from low to high elevations). Globally, animal endemism is higher where climate change velocity is lower, and this effect is stronger for sedentary amphibians than for mobile mammals and birds, suggesting that stable climates have promoted the persistence of sedentary species with small geographic ranges (Figure 6; Sandel et al. 2011).

Endemism has been an important concept in conservation, as manifested by efforts to identify hotspots of high numbers of species found nowhere else. In the most famous such analysis, Myers et al. (2000) defined global hotspots as regions with more than 1,500 endemic plant species and more than 75 percent loss of primary vegetation cover. The 25 hotspots thus identified make up only 1.5 percent of the world's terrestrial land surface but hold over one-third of the world's vertebrates in four major groups, as well as one-half to two-thirds of the plants and vertebrates on the IUCN Red List of Threatened Species. Nearly all hotspots are in the moist tropics or subtropics, contributing to their high overall diversity, and in the tropics they tend to be on islands or in islandlike mountain ranges, contributing to high endemism. Almost the only nontropical hotspots are found in the five mediterranean climate regions of the world, including the California Floristic Province. These five regions are also lower in vertebrate diversity than the other hotspots.

Global analyses of animal endemism generally find similar results to those of Myers et al. (2000), except where the mediterranean climate regions are concerned. For example, Rodrigues et al. (2004) used distributional maps for mammals, amphibians, turtles, tortoises, and endangered birds to identify the global regions with highest "irreplaceability," or numbers of species not found or protected anywhere else. The results highlighted many of the same tropical islands and mountain ranges identified by Myers et al. (2000) but not the five mediterranean climate regions. Likewise, Lamoreux et al. (2006) found that hotspots of amphibian, bird, mammal, and reptile endemism tended to coincide, but none of these concentrations occurred in the mediterranean regions. Nor did the mediterranean regions score as globally significant for total, endemic, or endangered birds (Orme et al. 2005)

Within the United States, including California, hotspots of endemism have been identified using a metric called rarity-weighted richness (Stein et al. 2000; CDFW 2003). A region is divided into equal-area polygons, within each of which the rarity-weighted richness is the sum of each species present in the polygon divided by the number of polygons occupied by that species. The output is a map showing high concentrations of narrowly distributed species (Figures 7, 8). The input data are often coarse and incomplete; in these examples, only Heritage Network–listed species are included, and their distributions are less than fully known. Also, the results may sometimes be dominated by small numbers of imperiled species with very tiny ranges; there is no single "correct" way to balance the contributions of number of species and range sizes in this type of analysis. Nonetheless, it provides a synoptic view of biodiversity that emphasizes endemism.

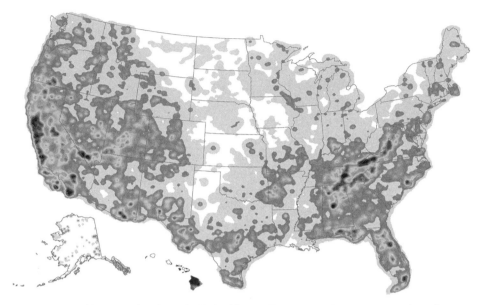

FIGURE 7. Hotspots of rarity in the United States. The rarity-weighted richness (RWR) analysis of critically imperiled and imperiled species shows concentrations of limited-range species, thus highlighting locations with species composition different from adjacent areas. The analysis points to locations that are essentially "irreplaceable" and present conservation opportunities found in few other places. (Source: NatureServe and its Natural Heritage member programs, July 2008. Produced by National Geographic Maps and NatureServe, December 2008.)

Within the United States, hotspots of rarity-weighted richness for Heritage Network–listed species (Figure 7) are the greater San Francisco Bay Area, part of coastal and interior Southern California, Death Valley, Hawaii, the Florida Panhandle, and the southern Appalachians. The first two reflect large numbers of plants threatened by both natural rarity and urbanization. Plants comprise 113 of 135 imperiled species in the San Francisco Bay Area and 72 of 81 in the Southern California hotspot (Stein et al. 2000). The Death Valley hotspot comprises fish, snails, insects, and plants confined to one or a few desert springs. Appalachia's imperiled species are mainly freshwater mussels, fish, and amphibians; Hawaii's include many birds; and Florida's include many reptiles and amphibians, although plants are also important in all three (Stein et al. 2000).

Within California, concentrations of rarity-weighted richness for nearly all groups of species (including mammals, birds, reptiles, and

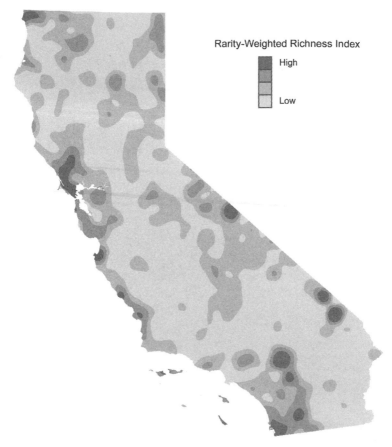

FIGURE 8a. Hotspots of rarity-weighted species richness in California. (a) For plants, particularly rich areas occur in the greater San Francisco Bay Area and western San Diego County, reflecting both ecological reality and human activity.

invertebrates) occur in the San Francisco and the Los Angeles–San Diego urban coastal regions (CDFW 2003). High levels of sampling effort and high degrees of endangerment may contribute to this pattern. For plants (Figure 8a), there are additional and presumably more natural hotspots in the Klamath-Siskiyous, Modoc Plateau, southeastern Sierra, Central Coast, and Mojave Desert. For amphibians (Figure 8b), the southern Sierra Nevada and Transverse Ranges stand out; and for fish (Figure 8c), the Modoc Plateau and Death Valley are especially rich in small-ranged species.

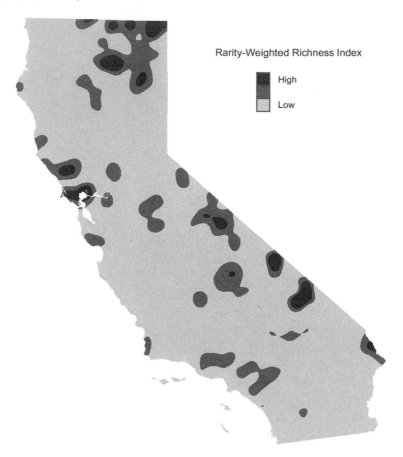

FIGURE 8b. For fish, important hotspots of restricted species include the Modoc region and the Mojave Desert.

ORIGINS OF ENDEMIC SPECIES

Paleoendemism and Neoendemism

As pointed out by Stebbins and Major (1965), species may be restricted to a narrow geographic area for two general reasons: either they were once widespread and are now confined to a subset of their former ranges (paleoendemism), or they evolved recently and have had inadequate time to spread (neoendemism). (Some authors consider a third category, "holoendemics," or species restricted by their habitat requirements.) Although humans have had enormous impacts on biological diversity, there are relatively few "anthropoendemics" whose

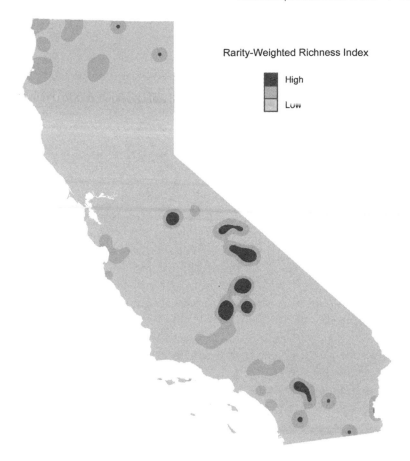

FIGURE 8c. For amphibians, the southern Sierra Nevada is a hotspot of rare species diversity. (Source: California Natural Diversity Database; CDFW 2003. Copyright California Department of Fish and Wildlife.)

small ranges are attributable to humans (Kruckeberg and Rabinowitz 1985).

Paleoendemics or relictual taxa may have fossil records far beyond their contemporary ranges, like ironwood (*Lyonothamnus*), found in western Nevada fossil beds but now confined to the Channel Islands. They often occur in discontinuous populations, like the coastal redwood (*Sequoia sempervirens*) and giant sequoia (*Sequoiadendron giganteum*), thought to be remnants of once-broader distributions. Usually their habitats are more benign than the surroundings, with higher-than-average summer rainfall and/or mild summer temperatures. Paleoendemics are also classically diagnosed by lacking close relatives nearby, since they represent old and shrinking lineages. The closest relatives of many Californian paleoendemics are found in eastern North America or eastern Asia (Stebbins and Major 1965; Raven and Axelrod 1978). By contrast, neoendemics belong to lineages undergoing recent speciation, so they may occur in complexes of closely related and adjacent species and may be poorly differentiated in morphology, genetics, and/or reproductive compatibility. Examples are discussed in Chapters 3 and 4.

The concepts of paleoendemism and neoendemism have also been applied to edaphic (soil) endemism. Paleoendemism is illustrated by species such as leather oak (*Quercus durata*), which is confined to serpentine soils but scattered across California and is believed to have once occurred widely on other soils but to have become restricted to serpentine through changes in the climate and/or competitive environment. Neoendemism is exemplified by species such as *Layia discoidea,* which appears to have arisen recently from a nonserpentine ancestor and to have been restricted to serpentine ever since it evolved (Baldwin 2005).

The distinction is not absolute, of course. Although *Lyonothamnus* is paleoendemic as a genus, its extant representative, *L. floribundus,* with its two distinctive subspecies, probably evolved recently on the Channel Islands (Erwin and Schorn 2000). The *Streptanthus glandulosus* complex appears to consist of a widespread species that became fragmented (similar to a paleoendemic), but some of its members have speciated recently as a result of their restriction to serpentine soil (Mayer et al. 1998). Although the prefixes *paleo-* and *neo-* imply differences in age, it should not be assumed that paleoendemic lineages are older unless (as is seldom true) this is actually tested.

It is usually assumed that neoendemism accounts for most of the wealth of endemism in California. It is certainly true that most of the studies of plant endemism in the state have focused on the evolutionary processes giving rise to new species in the region. As a background to the following chapters, the rest of this section briefly reviews modes of speciation by which new endemics evolve. The fundamental challenge is to understand how a single lineage can give rise to two (or more) descendant lineages that remain on separate evolutionary pathways rather than lose their integrity through gene flow. Geographic barriers, natural selection, hybridization, chromosomal rearrangements, and genetic architecture play roles that have been studied and debated for decades.

Geographic Speciation

One basic distinction is between geographic and ecological speciation, where the former implies a strong role for external barriers to gene flow as opposed to natural selection. In the classic model of gradual allopatric divergence, an ancestral lineage becomes subdivided by a new mountain range, water body, or similar obstacle. Sequential fragmentation may result in complexes of species with parallel patterns of genetic distance, morphological variation, and variable interfertility. Relatives that co-occur tend to have diverged longer ago than relatives that do not, suggesting that isolation promoted their initial divergence (Baldwin 2006). This is a classic and uncontroversial mode of speciation. It has been studied using the methods of biogeography, where the distributions of closely related taxa are interpreted in light of geologic and climatic events, and more recently using phylogeography, where the genetic patterns within lineages are similarly correlated to historical earth surface events. Biogeographic and phylogeographic evidence suggest that nearly all North American mountain chains are "suture zones," that is, places where plant and animal lineages have diverged and sometimes come into secondary contact (Swenson and Howard 2007).

Ecological Speciation

Natural selection in response to new ecological opportunities is central to ecological speciation. The most spectacular examples are the adaptive radiations that sometimes occur on newly formed islands or in islandlike habitats (e.g., Price 2008). Speciation into new niches also takes place in habitats already full of species, but formidable obstacles make its success unlikely (Levin 2004). Ecological speciation begins with a small lineage colonizing a habitat in which it is ill adapted. It is more likely to go extinct than to continue evolving because of its small population size, its low genetic variability, and the reduction in its potential population growth implied by the existence of strong selective pressures. Also standing in the way of successful adaptation are gene flow from populations in the ancestral habitat and correlations between adaptive and nonadaptive genetic traits. Even if the fledgling species becomes well adapted to its new environment, it risks being genetically swamped by its progenitor species until intrinsic reproductive isolation evolves, which may take longer than adaptation to the habitat. Hybrid origins and allopolyploidy help overcome these obstacles by creating immediate reproductive barriers between a newly adapted species and its ancestors; however, the new species may be ecologically outcompeted by its relatives unless it inhabits a novel niche (Levin 2004). Other pathways to reproductive isolation include strong selection against hybrids (Kay et al. 2011), selection that incidentally favors differences in reproductive traits (e.g., flowering phenology, floral morphology), and linkage or epistatic interactions between genes involved in adaptation and genes that confer reproductive isolation (Wu et al. 2007).

Progenitor-Derivative Speciation

Another basic distinction is whether speciation results in sister taxa with roughly equal initial population sizes, geographic ranges, and genetic diversity, as in classic gradual allopatric divergence, or whether it involves a small population budding off within the range of a widely distributed species. The latter case, called peripheral isolate formation or progenitor-derivative speciation, leads to a localized neoendemic species that is phylogenetically embedded in its ancestral species. The ancestor then becomes paraphyletic; that is, it does not include all descendants of a single common ancestor. Progenitor-derivative speciation appears common in plants (Grant 1981; Gottlieb 2003; Baldwin 2006). It is related to the classic idea of "catastrophic speciation," in which a sudden event causes a population decline in a widespread species, allowing a random chromosomal arrangement to become rapidly fixed in a small population that evolves into the derivative species. Chromosomal alterations, novel habitats, breeding system changes, and adaptive morphological differences may contribute to reproductive isolation between progenitor and derivative species. Changes appear to be moderate and to involve a small number of genes, and overall genetic similarity between progenitor and descendant remains high (Gottlieb 2003).

Hybrid and Polyploid Speciation

The origin of new species through hybridization and/or changes in chromosome number is sometimes detected in animals but is central to evolution in plants. In his classic *Plant Speciation* (1981), Grant argued that patterns of relatedness among plant lineages are so often network-like, due to both modern and ancient hybridization events, that the standard view of evolution as a branching "tree of life" may not apply.

Hybridization generates adaptive potential because it increases genome size and allows various duplicate genes to be turned on or off in hybrid progeny. Hybridization between species with different chromosome numbers or structures initially produces progeny of low fertility because of chromosomal incompatibility, but chromosomal doubling or other rearrangements may restore fertility, in addition to causing reproductive isolation between hybrid and parental lineages. The resulting allopolyploid hybrids are thought to be particularly capable of rapid evolution because they have the full chromosomal complement of both parents. Hybridization between parental species with the same chromosome numbers and structure produces hybrids with the same chromosome number as the parents. These homoploids face lower initial barriers to fertility than allopolyploid hybrids but are more likely to be genetically swamped by the parental lineage. For reproductive isolation to develop in homoploid hybrids, within-chromosome rearrangements and geographic or ecological isolation may be necessary (Rieseberg 2006). Both allopolyploid and homoploid hybrids may exceed the parental species in the values of quantitative traits, and such "transgressive hybridization" may facilitate the invasion of new niche space (Grant 1981; Rieseberg 2006). Autopolyploids are products of within-species gene duplication rather than hybridization. Like allopolyploids, they enjoy the adaptive benefits of larger genome sizes (Soltis et al. 2004).

Recent molecular studies have confirmed the conclusions of classic authors that many plant lineages are of polyploid origin. Individual polyploid "species" may arise multiple times. Autopolyploidy and genome-wide duplication events are more common relative to allopolyploidy than was once believed. Rapid chromosomal rearrangements, genomic downsizing, and changes in gene expression following polyploid origins are beginning to be studied, as are the relationships of these genomic changes to pollination, reproductive biology, and ecological traits. A growing number of comparative studies illustrate the potential for polyploid hybrids to have new, broader, or more finely partitioned niches than their ancestors. Polyploid lineages may diverge ecologically or reproductively through the loss or silencing of alternative duplicate genes, contributing to their evolutionary dynamism (Soltis et al. 2004).

Hybridization and polyploidy have been credited with important roles in the rapid evolution of Californian flora (Stebbins and Major 1965). A common pattern in the California flora is for close relatives to hybridize intermittently, perhaps in certain zones, yet to remain distinct over their core geographic distributions. This pattern, seen in *Arctostaphylos, Chorizanthe, Eriogonum, Monardella,* and other taxa, is informally known to botanists as the "California pattern" (Skinner et al. 1995).

TRAITS OF ENDEMIC SPECIES

Rare species are sometimes found to have lower genetic variability, higher rates of selfing, lower reproductive investment, poorer dispersal, higher susceptibility to natural enemies, or less competitive ability than common ones (Kruckeberg and Rabinowitz 1985; Lavergne et al. 2004). These traits are interpreted as factors that may cause rarity, that is, prevent species from achieving higher range sizes or abundances. Other studies find that rare species inhabit less competitive (e.g., rockier) or more benign (e.g., rainier, less seasonal, less fragmented) environments than their common relatives (Lavergne et al. 2004; Harrison et al. 2008); such extrinsic differences may be interpreted as factors that have helped species persist, given that they are rare for other reasons. Finally, some traits such as higher inbreeding and lower genetic variability could be interpreted as either causes or consequences of rarity. Not many strong generalizations have arisen from the literature on the biology of rarity, and the usual conclusion is that rarity is too complex to have a single cause.

To understand high endemism within a geographic region such as California, it would be interesting to ask whether endemics have any consistent attributes, either intrinsic or environmental, that explain their diversity relative to other taxa and other regions. For example, in the Cape flora of South Africa, it has been proposed that plant adaptations to nutrient-poor soils, including fine, oil-rich, flammable foliage and ant-dispersed seeds, produce high rates of speciation by conferring short generation times and low dispersal, thereby leading to exceptional diversity (Cowling et al. 1996). In California, Wells (1969) proposed that the diversification of *Arctostaphylos* and *Ceanothus* is linked to the loss of resprouting in these two shrub genera, which regenerate by seeds after fire and thus have shorter generation times than their resprouting relatives. The annual life form is a widespread adaptation (or preadaptation) to California's summer drought that may similarly facilitate rapid speciation (Raven and Axelrod 1978). There are now formal phylogenetic methods for testing the relationship between such possible "key innovations" and rates of diversification, but they have not yet been applied to California endemism. However, specialization to serpentine in 23 genera of California plants was shown to be the opposite of a key innovation; when lineages become endemic to serpentine, their diversification rates decline (Anacker et al. 2011).

. . .

Endemism is biologically interesting as long as the geographic unit being studied is meaningful in terms of processes that create and maintain diversity. Californian endemism is best studied at the scale of the California Floristic Province, a natural biotic unit defined by a flora with a shared evolutionary history, but this book also considers state-level endemism due to data constraints. Taxonomic scale also influences the study of endemism. This book focuses on full species because some groups of organisms are much more extensively split than others into units below the species level, which by definition have higher levels of endemism.

Species richness is highest in parts of the world where water and solar energy are abundant. The richness of endemic species is also high on islands and in historically stable climates. The mediterranean regions of the world are unusual in being rich in plant endemism, yet not as obviously rich in animal endemism, whereas other global hotspots (nearly all of which are tropical) do not show this disparity.

Endemics arise through the shrinkage of widespread geographic ranges (paleoendemism) and the evolution of new species (neoendemism). Pathways to neoendemism include allopatric divergence, ecological speciation, peripheral isolate formation, and hybridization, all of which are well known in the Californian flora. There is not yet a predictive understanding of the traits that make lineages likely to diversify in particular environments such as California's.

A Brief History of California

Plant and animal diversity in California are clearly linked to the rich complexity of the contemporary landscape, including its rugged topography, (see Figure 1), many climatic zones, varied geologic substrates, and resulting tapestry of vegetation types. This ecological variety has been well described in many places; see, for example, Barbour et al., *Terrestrial Vegetation of California* (2007), for plant communities and vegetation; and CDFW, *Atlas of the Biodiversity of California* (2003), for a pictorial overview of plant and animal diversity in relation to the landscape. This chapter presents a brief overview of how California's ecological landscape evolved (for other accounts, see Edwards 2004; Minnich 2007; Millar 2012). The monograph by P. H. Raven and D. I. Axelrod, *The Origins and Relationships of the Californian Flora* (1978), is central to this discussion. Raven and Axelrod's account is exceptional in its sweeping, yet detailed view of the history of a regional flora. Although it has been critiqued on various counts, it has not yet been replaced. This chapter uses the Raven and Axelrod story as an essential starting point. Chapter 3 reconsiders the classic story in light of more recent ideas and evidence.

GEOLOGIC HISTORY

During much of evolutionary history, ocean existed where California is today, and the coastline has gradually grown westward through a series of plate tectonic events (Figure 9; Table 2). Just over 200 million years

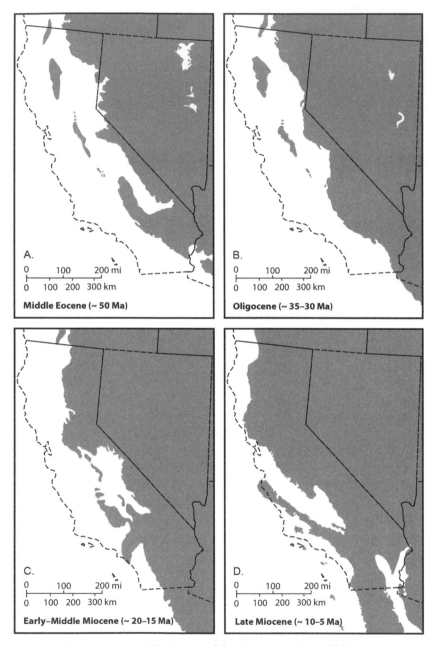

FIGURE 9. Paleogeography of California in (a) mid-Eocene, 50 Ma; (b) Oligocene, 35 Ma; (c) mid-Miocene, 20 Ma; and (d) late Miocene, 10 Ma. (Source: Ron Blakey, Colorado Plateau Geosystems)

TABLE 2   THE GEOLOGIC TIME SCALE
(IN MILLIONS OF YEARS)

Era/Period	Ma
Cretaceous	
Early Cretaceous	146–112
Late Cretaceous	112–65
Tertiary	
Paleocene	65–56
Eocene	56–34
Oligocene	34–23
Miocene	23–5.3
Pliocene	5.3–2.6
Quaternary	
Pleistocene	2.6–0.012
Holocene	0.012–present

NOTE: The time scale has changed over recent decades; for example, the period 12–5.3 Ma was considered Early Pliocene at the time of Raven and Axelrod (1978) but has since been reassigned to Late Miocene.

ago (Ma), the supercontinent Pangaea broke up and North America began colliding at its western edge with smaller oceanic plates known as terranes, causing subduction (movement of one plate beneath another) and accretion (addition of one plate to another). By 140 million years ago, the edge of the continent had reached the present-day location of the Sierra Nevada and its western foothills. Then the same processes shifted farther westward and built the Coast Ranges. Subduction ceased around 28 million years ago when the zone where it was occurring collided with the East Pacific Rise (the midocean trench or spreading center). The subduction zone gave way to a mostly horizontally moving plate boundary fault, namely, the complex of northwest-to-southeast faults known as the San Andreas system, along which the motion is relatively northwest on the west side and southeast on the east side. This change in plate motion ultimately led to the rise of the Coast, Peninsular, and Transverse Ranges.

The Sierras were first uplifted beginning about 80 million years ago and subsequently eroded into an undulating plain that rose gradually 50 million years ago to Tibet-like heights in east-central Nevada. The present Sierra Nevada includes remnants of this old surface as well as younger granitic rocks that intruded as subduction occurred in the Coast Ranges. A second period of more rapid uplift of the Sierra Nevada began around 3 million years ago. The Klamaths either represent the northern end of the Sierra offset westward by a fault 130 million years ago or are the remains of oceanic terranes lying west of the northern continuation of the Sierra Nevada. During much of the Eocene, 56 to 34 million years ago, the Klamath region was an island with a stable land surface that eroded in a tropical climate, and much of this land surface still exists. The uplift of the Coast Ranges began in Southern California around 30 million years ago and migrated northward with the north end of the San Andreas fault and the Mendocino Triple Junction, where it currently continues. The Coast Ranges began as an offshore submerged subduction complex but were fully joined to the continent by 5 million years ago. The east-west Transverse Ranges arose at around that time because of a bend in the San Andreas fault that resulted in compression during the northward movement of the Pacific plate along the fault.

The Central Valley is one of the largest and flattest valleys in the world. It is believed to have been created when a large slab of oceanic plate was thrust over the North American continental margin during the collison of North America with a west-dipping subduction zone. A new east-dipping subduction zone formed west of this slab in the modern Coast Ranges, leaving a broad gap in between. It lay beneath ocean until about 5 to 15 million years ago depending on location. Between about 5 and 2 million years ago, mountain uplift created a marine embayment in the Central Valley encircled by mountains and draining to the ocean from the southern valley near Monterey Bay. Continued mountain uplift later blocked this seaway, and by about 600,000 years ago the Central Valley was a freshwater basin draining through the region of present-day San Francisco Bay (Harden 2004).

Deserts in southeasternmost California contain ancient continental rocks that attest to their having been part of ancient North America rather than accreting onto the edge, as did most of the west coast. The deserts also harbor more extensive fossils of ancient terrestrial life than the rest of the state, including Eocene terrestrial vertebrates. However, their history as deserts is quite recent (see below).

The California Channel Islands are seafloor ridges, transported northward, rotated, and uplifted by movement on the San Andreas fault beginning 28 million years ago. Although the four smaller islands were inundated at various times in the Pleistocene, beginning 1.8 million years ago, parts of the four largest islands have been continuously exposed for at least 600,000 years, and it is even possible that they had ancient (Miocene) connections to the continent. During the lower sea levels of the Pleistocene, the four northern islands were united into the superisland of Santa Rosa lying only 6 kilometers from the mainland. Santa Cruz, the closest island today, is now 30 kilometers from shore.

CLIMATIC HISTORY

The global climate has generally cooled and dried over the past 50 million years, and plate tectonics have played an important role by altering oceanic circulation and atmospheric composition. Major climate-changing events have included the opening and widening of Antarctic ocean passageways, the uplift of the Himalayas and consequent decline in atmospheric CO_2 due to chemical weathering, and the arrival from the west of Panama and closure of the Central American seaway. Superimposed on these longer-term trends are oscillations on the order of tens to a few hundreds of thousands of years, caused by variation in the eccentricity, obliquity, and precession of the earth's orbit around the sun (Milankovitch cycles). The interaction of these forces is complex. In particular, it appears that the tectonically driven changes have increased the sensitivity of the climate system to orbital forcing, leading to increasingly rapid and extreme climatic fluctuations toward the present day (Zachos et al. 2001).

The earth's transition "from greenhouse to icehouse" has been reconstructed mainly from the carbon and oxygen isotopic composition of foraminiferan shells recovered from Antarctic deep-sea drilling (Figure 10). Warming trends from the mid-Paleocene (59 Ma) to early Eocene (52–50 Ma) produced the Eocene climatic optimum, a time when much of the earth experienced climates resembling today's wet tropics, except for caps of temperate climate at the poles. This was followed by a long period of cooling, leading to the formation of Antarctic ice sheets by the early Oligocene (34 Ma). Moderate warming from the late Oligocene (26 Ma) to the middle Miocene (15 Ma) reduced the extent of oceanic ice, although temperatures never regained their Eocene levels. Cooling resumed from the middle Miocene to the middle Pliocene (6 Ma). The Northern Hemisphere Glaciation, beginning 3.2 million years ago, marked the beginning of extreme oscillations, with more than 20 relatively long glacial periods interrupted by shorter interglacials (Figure 10; Zachos et al. 2001). The present interglacial period began at the Pleistocene-Holocene boundary 12,000 years ago and is expected to end with another ice age unless disrupted by anthropogenic additions of greenhouse gases.

Today's five mediterranean climate regions with their characteristic winter rainfall and summer drought are found on west coasts between

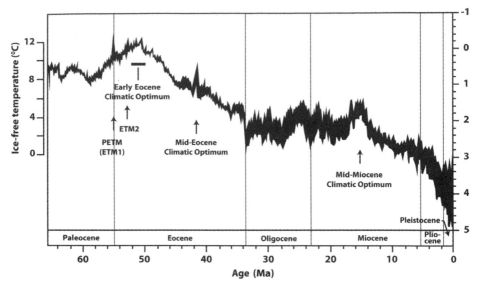

FIGURE 10. Global relative temperatures from the Paleocene to the present, based on oxygen isotope measurements from deep-sea sediments and ice cores. (Based on data from J. Zachos; see also Zachos et al. 2008)

roughly 30° and 42°; latitude (Figures 2, 3). In these locations, the jet stream brings winter storm systems from maritime rather than interior sources, leading to cool, rainy winters instead of cold, snowy ones. The high-pressure systems that create the world's major deserts (from 23° to 30° latitude) shift poleward in summer with the earth's tilt, creating the desertlike summer drought. Upwellings of cold deep-ocean water along the coast, which produce a marine layer of cool air capped by warmer inland air, are also a key ingredient of both the summer drought and the relatively gentle winter weather. Timing and duration of the summer droughts vary considerably among the five world regions, with California's being among the longest and driest (Dallman 1998).

Because of its relevance to plant and animal evolution, the history of the mediterranean climate is of great interest. In today's mediterranean zones, tropical-like climates began to give way to more seasonal ones as many as 40 million years ago. Precipitation began to peak in the winter by the middle Miocene. However, the fully mediterranean climate with its near-complete summer drought emerged only after the onset of the Pleistocene brought the development of cold offshore currents. While glaciation in the Northern Hemisphere was well under way 2.7 million years ago, the modern ocean current system was not in place until about

1.5 million years ago (Ravelo and Wara 2004). The uplift of the state's major mountain ranges in the past 5 million years contributed to increasingly steep internal climate gradients. Based on plant fossil evidence, Raven and Axelrod (1978) argued that some summer rainfall persisted in California until one million years ago, but this has yet to be corroborated with geophysical evidence.

Another important question is how severely the climate fluctuated during glacial-interglacial cycles. The conventional wisdom, largely from plant-based evidence (see later sections of this chapter), is that the climate was colder and rainier but remained mediterranean during glacial periods. However, isotopes indicate that during the coldest parts of the last several glacial periods, expanded oceanic ice sheets blocked the cold oceanic current system, producing warmer and rainier conditions in California resembling a prolonged El Niño. Fossil pollen indicates that while this advance warming speeded the recovery of the interglacial vegetation, it caused declines in abundance of the fog-dependent coastal redwoods (Herbert et al. 2001).

FLORISTIC HISTORY

Origin of the Flora According to Raven and Axelrod

Building on their decades of research in plant evolutionary biology and paleobotany, respectively, Raven and Axelrod (1978) described the biotic history of the California Floristic Province beginning with the Eocene, when many modern plant families diversified and most of the world had a warm, wet, essentially tropical climate. The Sierra was a low coastal mountain range, and the Klamaths were offshore islands. The Coast Ranges had not yet emerged. Many parts of present temperate North America and Eurasia were covered in a tropical rainforest flora containing families and species that are now largely extinct north of the humid subtropics, including laurels (Lauraceae) and palms (Palmae), among others.

However, the more northerly and mountainous interior regions of this Eocene world contained what earlier authors had called the Arcto-Tertiary Geoflora, a rich mixture of trees, shrubs, and herbs whose descendants are now found in the temperate forests of East Asia and eastern North America. This flora has sometimes been described as resembling a modern redwood forest, although with many more species of both angiosperms and conifers. It included the ancestors of today's Californian coastal forests, which were found north of 44° latitude, as well as the ancestors of the drier montane Sierran forests, which were then found farther south. Both of these elements shifted coastward as cooling and drying accelerated in the Oligocene to Miocene, and many of the warm tropical forest elements became extinct. As revealed by fossil assemblages, these forests were enriched by the co-occurrence of species that are now segregated by elevation and habitat. Five million years ago, the mediterranean climate was clearly developing, but fossil floras indicate a less seasonal climate; the wide occurrence of *Abies* and *Picea* suggests cooler summers, while the presence of now-extinct *Persea, Castanea,* and *Ulmus* suggest wetter summers. Late Pliocene decreases in summer rainfall increasingly restricted conifers to high elevations, and cooling temperatures restricted broad-leaved evergreens to low elevations. Pleistocene vegetation was essentially modern, except that it was shifted downward in elevation and latitude compared to the present; conifers were more widespread due to cooler summers, and hardwoods survived in mild coastal climates. During the early Holocene warm period 8,000 to 4,000 years ago (called the "Xerothermic" by early authors), the remnants of the Arcto-Tertiary flora retreated toward their present coastal, riparian, and higher-elevation refuges.

Together with other authors, Raven and Axelrod (1978) considered California one of the most important areas for survival of the Arcto-Tertiary flora, second in the United States only to the considerably richer forests of Appalachia. The refuge existed because the climate remained consistently equable during and since the Tertiary, without widespread glaciation or extreme aridity. Within California the most significant Arcto-Tertiary refuge is thought to be the Klamath-Siskiyou region, where the greater total and summer rainfall, milder winters, and cooler summers amount to a climate resembling that of the Miocene and Pliocene. Elsewhere, Arcto-Tertiary forests became restricted to small patches during the mid-Holocene warm period, and monodominant stands of species such as *Pinus jeffreyi, Pinus ponderosa,* and *Pseudotsuga menziesii* developed in recent millennia through the progressive loss of other species.

Today's Arcto-Tertiary flora, as defined by Raven and Axelrod (1978), comprises just over half the species in the California Floristic Province and is the source of most of its paleoendemics. Some Arcto-Tertiary taxa that meet the standards for very strict paleoendemism, such as having their closest relatives outside of western North America, are *Quercus sadleriana, Picea breweri, Berberis nervosa,* and *Pinus albicaulis.* Paleoendemics in a slightly broader sense, such as plants having few close relatives in western North America, include *Chrysolepis chrysophylla, Taxus brevifolia, Torreya californica, Calycanthus occidentalis, Dirca occidentalis, Lithocarpus densiflorus,* and *Acer circinatum.* Neoendemics in California are not normally thought of as being of Arcto-Tertiary origin. However, Raven and Axelrod (1978: 16) listed 50 Arcto-Tertiary genera that have undergone significant speciation in California (e.g., *Allium, Aster, Bromus, Calochortus, Delphinium, Iris, Lomatium, Lupinus, Silene, Viola*). Together, these genera comprise 645 taxa, of which 253 are strictly endemic to California (see Chapter 3 for further analysis).

As the climate became drier in the mid- to late Eocene and the Arcto-Tertiary forests shifted coastward, a more drought-adapted flora began to expand northward into western North America. Fossil floras containing sclerophylls (species with hard, thick, drought-adapted leaves), such as species of *Quercus, Arbutus, Pinus,* and various laurels (Lauraceae), have been found in deposits as old as 50 million years. Axelrod named this broad assemblage the Madro-Tertiary Geoflora, based on a resemblance to the flora of today's Sierra Madre of northern Mexico. Axelrod speculated that this flora might have its ancient roots in Mediterranean Europe and/or in localized dry habitats such as rocky south-facing slopes in low-latitude North America. Some members, such as *Arbutus, Cupressus, Erodium, Lavatera, Quercus durata,* and *Q. berberidifolia,* were noted to have close relatives in the Mediterranean Basin; Raven and Axelrod termed these taxa "Madrean-Tethyan" after the ancient Tethys Sea that separated Laurasia from Gondwana. By the Miocene, the rich subtropical semiarid Madro-Tertiary (or Madrean) flora dominated interior Southern California. Besides the above taxa, it included Palmae, *Lyonothamnus, Ceanothus, Arctostaphylos, Heteromeles,* and *Rhus.* Continued drying in the Pliocene caused Madrean chaparral to expand farther into the Sierra foothills and coastal California and to lose some of its more tropical elements such as *Acacia* sensu lato. Raven and Axelrod visualized the Madrean flora meeting the Arcto-Tertiary flora at relatively abrupt boundaries along climatic gradients. These boundaries fluctuated during Pleistocene climatic cycles and largely arrived at their present configuration during the early Holocene warm period.

Raven and Axelrod (1978) believed about one-third of species in the California Floristic Province were of Madrean origin. The majority of the neoendemics belong to Madrean genera that radiated extensively in the province and are almost completely endemic at the genus level (e.g., *Clarkia, Hesperolinon, Lasthenia, Mimulus, Phacelia*). The neoendemic genera are believed to have undergone most of their diversification in the late Pliocene and the Pleistocene as the mountains rose and the climate became fully mediterranean. In the view of Raven and Axelrod (1978) and other classic authors (e.g., Stebbins and Major 1965), much of the speciation was stimulated by the climate-driven advances and retreats of Madrean vegetation across rugged landscapes, which created many opportunities for fragmentation, divergence, reproductive isolation, and/or subsequent hybridization. Preadaptation to summer drought and fire helped to determine which genera thrived and speciated in the new climate. Most annual herbs, and most shrubs that obligately recruit by seed after fire, are Madrean. There are also Madrean paleoendemics in the mountains of Southern California, representing wetter elements of the Madro-Tertiary flora that survived under occasional summer rainfall.

Raven and Axelrod also identified a second group of drought-adapted species that they termed the "warm temperate desert" element of the flora, which they thought moved into the California Floristic Province from the south during the mid-Holocene warm period and colonized the interior Coast Ranges and southern Central Valley. This group comprises 604 species, or 13.5 percent of the flora of the province. Raven and Axelrod (1978) estimated there were 44 endemics in the Central Valley, most of which they considered young (dating to the mid-Holocene warm period) and of desert origin.

Desert floras are relatively poor in endemic species because of the recency of the desert climate, according to Raven and Axelrod (1978). The Mojave and Great Basin Deserts were largely covered by pinyon-juniper woodlands during the Pleistocene. The Sonoran Desert was subtropical woodland. Having undergone less cooling and having retained moderate summer rainfall, the Sonoran Desert contains more relictual subtropical taxa. Raven and Axelrod (1978) enumerated a total of 102 genera and 935 species in the Californian deserts, including 9 species endemic to the Great Basin, 44 to the Inyo region that includes the White Mountains and Death Valley, 22 to the Mojave, and 8 to the Sonoran (Colorado) Desert.

Raven and Axelrod (1978) cited the geography of Californian endemism in support of their conclusions, relying largely on an analysis by Stebbins and Major (1965). Especially high concentrations of neoendemics were found in "intermediately" warm and dry mediterranean-type vegetation, particularly in coastal Southern California, the Sierran foot-hills, and the central Coast Ranges. Paleoendemics were found to be most prevalent in the Klamath-Siskiyou, the northern Coast Ranges, the Channel Islands (as also described in Raven 1965), and northern Baja California. Scarcity of both neo- and paleoendemics was noted in the climatically youthful Central Valley, deserts, and high Sierra.

In summary, Raven and Axelrod portrayed California's floristic history as a progressive shift from a largely mesic tropical and warm-temperate flora to a modern flora with a much more arid-adapted character, with new species arising as climatic oscillations across the rugged landscape produced a constant interplay between two distinctive assemblages. The generally equable climate of California enabled the survival of many mesic Arcto-Tertiary relicts. The cycles of cool/moist to warm/dry climates since the Pliocene triggered outbursts of speciation, mostly among the southerly Madro-Tertiary component of the flora, leading to especially high neoendemism within the fully mediterranean climates where the modern vegetation is chaparral.

Raven and Axelrod's (1978) account is remarkable for its attention to detail. Their sweeping historical analysis is complemented by attention to the numbers of species belonging to each geographic region, life form, and biogeographic origin. Tables in their monograph give the identities of the taxa interpreted as having different biogeographic affinities. For virtually no other large region in the world is there such a comprehensive attempt to link the identities of modern species to their places in a broad account of biogeographic history.

Critiques of the Classic Story

Scientific progress leaves every ambitious accomplishment open to reconsideration. Perhaps the most outdated aspect of the Raven-Axelrod analysis is its reliance on the geoflora concept. Geofloras were envisioned in the early and mid-twentieth century as widespread and long-lasting assemblages that formed in the Tertiary and remained constant in their ecological requirements, changing little through either trait evolution or differential migration and moving as a unit in response to climate change. This is clearly out of step with the modern view that emphasizes the individualistic nature of species range shifts, the recent assembly of modern communities from disparate ancestry, and the significance of adaptation as well as migration in response to climate change (e.g., Davis and Shaw 2001). The existence of an Arcto-Tertiary

Geoflora as classically defined was disputed by Wolfe (1978) on the basis of Eocene fossil floras from the Arctic that were predominantly broadleaf evergreen rather than deciduous. However, the notion of a north-temperate deciduous flora at Arctic latitudes in the Tertiary has been borne out in more recent analyses (e.g., Basinger et al. 1994; Brown 1994). More generally, it could be argued that the basic Raven and Axelrod story of the formation of the Californian flora, through range shifts and evolution in contrasting northerly mesic-adapted and southerly drought-adapted assemblages, could still be valid even if the shifts occurred in a less unitary fashion than these authors envisioned (Ackerly 2009).

Modern authors generally substitute more nuanced terms for the old geoflora names. This book follows Ackerly (2009) in using "north-temperate" in place of Arcto-Tertiary and "subtropical semiarid" in place of Madro-Tertiary, except when specifically citing Raven and Axelrod.

Terrestrial plant fossils are uncommon in California, and Axelrod's paleobotanical conclusions involved much interpolation from scarce data. In keeping with the geoflora concept and its principle that species have evolved little, Axelrod employed the assumption that a now-fossilized plant inhabited a climate much like that of its closest living relative. This method has been criticized for ignoring evolution and within-taxon diversity, although it may have some validity if large enough suites of species are used (Basinger et al. 1994). An alternative approach to inferring ancient climates is to use the physiognomic traits of entire fossil assemblages; for example, the proportion of plants with entire (smooth-margined) versus toothed and lobed leaves is strongly correlated with mean annual temperature in climates with year-round rainfall (Wolfe 1978). Another problem was that Axelrod assigned fossil taxa to modern genera using subjective matching of easily visible traits, such as leaf outline and major venation (Edwards 2004); newer approaches to fossil identification emphasize venation patterns and the microscopic examination of epidermal anatomy (Ellis et al. 2009).

More recently, stable isotopes have increasingly allowed paleoclimates to be reconstructed independently of plant fossils. One of the most important findings has been that "the Ice Age" was not a single cold event, as was once believed, nor was the early Holocene warm period an aberrant extreme. Rather, the Pleistocene consisted of many glacial cycles, interspersed by periods that often reached temperatures as warm as the early Holocene (Millar 1996). Complex changes in Californian plant distributions occurred during glacial-interglacial

cycles, with both herbs and hardwoods tending to expand as the gla-
cials ended (Edwards 2004).

Another problematic issue is that plants were designated as Arcto- or
Madro-Tertiary by subjective methods that relied on species traits and
contemporary distributions as well as fossil evidence. Thus groups such
as *Brodiaea* and its relatives, having many species in the province and
few outside it, were designated Madro-Tertiary because "the degree of
radiation . . . in the California Floristic Province suggests a relatively
great antiquity for that group in Madrean vegetation" (Raven and
Axelrod 1978: 50). In other words, they are of Madrean origin because
they are diverse in Madrean vegetation. While such inferences could
well be correct, corroboration from independent evidence would lead to
stronger inferences. Moreover, the concept of biogeographic "origins"
has its problematic aspects, since (for example) a genus may be inferred
to have arisen in one region or climate but its family in another. For
adherents of the geoflora concept, the Eocene is regarded as the period
when modern lineages acquired the traits defining their climatic niches,
an assumption that has never been tested. Ackerly (2009) notes that
Raven and Axelrod seem to equate the geographic region in which
greatest diversification occurred with the region of origin, although the
two need not be the same. An alternative approach would be to focus
on traits, which unlike lineages originated at specific times; phylogenetic methods can be used to ask in what regions and climates a trait
arose and how it affected the subsequent spread and diversification of a
lineage (Ackerly 2009).

Raven and Axelrod's classification of species in terms of biogeographic
origin has been examined in a number of recent analyses. Ackerly (2003)
found that woody species classified as Madro-Tertiary had larger seeds
and lower specific leaf area (i.e., thicker leaves) than those classified as
Arcto-Tertiary. Harrison and Grace (2007) and Ackerly (2009) found
that the geographic distribution of the groups conforms as expected to
climatic patterns within the state; Raven and Axelrod's Arcto-Tertiary
species are most numerous in the rainy and mountainous north, Madro-
Tertiary species (including those "strongly associated with the California
Floristic Province," many of which are endemics) were most abundant
in the Coast Ranges and Sierra Nevada foothills, and desert species
tend to be found in the driest parts of the California Floristic Province.
Damschen et al. (2010) found that over a six-decade period (1949–2007),
as mean temperatures in the Siskiyou region increased by 2°C, the abundance of north-temperate (Arcto-Tertiary) forest herbs declined relative

to other species. All these analyses bear out, to some extent, the existence of genuine differences between the lineages subjectively identified by Raven and Axelrod as Arcto-Tertiary and Madro-Tertiary.

Phylogenetic analyses of molecular data are a vast new source of evidence on evolutionary processes that were unavailable to Raven and Axelrod (1978). Chapter 3 considers what is now known about Californian plant endemism in light of this and other new evidence.

. . .

The Californian landscape has come into existence over the past 200 million years, and the modern climate and flora have developed during around 50 million years of global cooling and drying. Rapid geologic and climatic changes within the past 5 million years have left an especially strong imprint on the present-day biota. In the historic account by Raven and Axelrod (1978), the Californian flora arose from the interplay of two main sources: a northerly, temperate assemblage (Arcto-Tertiary) and a subtropical, arid-adapted (Madro-Tertiary) assemblage.

The north-temperate assemblage is thought to have given rise to many paleoendemics with relatives in eastern North America or eastern Asia. They include many broad-leaved deciduous trees and shrubs, conifers, and perennial herbs, and are most prevalent in cool and wet environments. The subtropical seminarid assemblage is considered to have given rise to the majority of neoendemics in the California Floristic Province. These species are often evergreen shrubs, geophytes, or annuals. Their centers of diversity are in mediterranean-type chaparral and coastal scrub habitats. Most neoendemics are thought to have arisen in the past 5 million years or less, since the climate became fully dry in the summer, although recognizable relatives have been found in 20- to 30-million-year-old fossil deposits. However, many aspects of this story have been reexamined in recent decades and many new details added. The next chapter examines both old and new evidence on the origins of Californian plant endemism and asks how well the classic story holds up.

3

Plant Endemism in California

Patterns and Causes

California's endemic plants have long been the objects of attention from evolutionary biologists, beginning with classic mid-twentieth-century biosystematics studies of *Ceanothus, Layia,* and other endemic-rich groups that played an influential role in the modern synthesis of evolution and genetics (Stebbins 1950; see also Smocovitis 1997). Early quantifications of neoendemism and paleoendemism (Stebbins and Major 1965), insular endemism (Raven 1967), and edaphic endemism (Kruckeberg 1954, 1984) in Californian plants gave rise to basic concepts about plant endemism that remain in wide use today. As molecular and phylogenetic techniques have become more prominent, Californian endemic plants have been examined in studies of diversification (e.g., *Clarkia,* Gottlieb 2003; *Mimulus,* Beardsley et al. 2004; *Layia,* Baldwin 2006; *Hesperolinon,* Springer 2009; *Ceanothus,* Burge et al. 2011), adaptation to novel habitats (e.g. *Mimulus,* Wu et al. 2007; *Linanthus,* Kay et al. 2011), genetic consequences of narrow relictual distributions (e.g., *Pinus torreyana,* Ledig and Conkle 1983; *Pinus radiata,* Millar 1999), and genetic and evolutionary processes in island populations (e.g., chaparral shrubs, Bowen and Van Vuren 1997; *Camissionia* and *Cryptantha,* Helenurm and Hall 2005; *Deinandra,* Baldwin 2007).

This chapter evaluates classic and modern evidence on Californian plant endemism: geographic patterns of species and genetic diversity within the state, genetic and evolutionary analyses, fossil evidence, and comparisons between the Californian flora and flora in the other mediterranean climate regions. Together, this makes it possible to reevaluate the classic story told in the previous chapter and to arrive at some conclusions about the relative importance of modern climate, climatic history, geographic barriers, and other factors.

ESTIMATING PLANT ENDEMISM

Raven and Axelrod's (1978) figures for Californian endemism are comprehensive and remain widely cited. For the state, they enumerated 1,517 endemics (30%) out of 5,046 total species and 26 (3%) endemics out of 878 genera. For the Floristic Province, the corresponding figures are 2,125 (48%) endemics out of 4,452 species and 50 (6%) endemics of 795 genera. They noted that southwestern Oregon contributes 40 endemics and 90 total species, and northwestern Baja 107 endemics and 227 total species to the Floristic Province that are not found in the state. Other sources have cited state endemism as 1,416 species (30%) (Stein et al. 2000) and Floristic Province endemism as 2,124 species (60.9%) (Conservation International 2011), with reasons for the discrepancies being unclear. A recent analysis of collection and literature data finds 106 species endemic to the mediterranean climate coastal corridor of Baja California, as well as another 66 species endemic to the Sierra San Pedro Mártir at the northern end of the peninsula (Riemann and Ezcurra 2007).

For the state of California, new endemism figures accounting for recent taxonomic and distributional revisions are given in Table 3. Note that the figures in Table 3 under the heading "State Endemics Found within the California Floristic Province" are not the same as, nor are they comparable to, "all species endemic to the California Floristic Province." Rather, they represent species that are restricted to the state and that occur within its Floristic Province regions. A preliminary updated list of Floristic Province endemic plants, including those in Oregon and Baja California, is given in the Appendix.

California's endemic plant richness easily exceeds that of any other U.S. state (e.g., Texas, 251 species, 6%; Florida, 155 species, 5%), although Hawaii's proportional endemism is higher (1,048 species, 87%) (Stein et al. 2000). On a global level, the California Floristic Province is similar to Japan (1,371, 34%) and New Zealand (1,618, 81%). Although not rich in endemic genera, it exceeds Hawaii, Japan, and New Zealand (31 [12%], 17 [2%], 39 [10%], respectively) (Raven and Axelrod 1978).

When ecoregional endemism is considered, California is less impressive because it is divided among 13 ecoregions. The Colorado Plateau

TABLE 3   NUMBERS OF PLANTS STRICTLY ENDEMIC TO THE STATE OF CALIFORNIA

	Genera		Species		All Named Taxa	
	Total	*Endemic*	*Total*	*Endemic*	*Total*	*Endemic*
	992	59 (6%)	5265	1,455 (28%)	6,506	2,264 (34%)
State endemics by life-form						
Woody			1,176	297 (26%)	1,620	593 (37%)
Perennial herb			1,733	476 (27%)	2,091	704 (34%)
Annual herb			1,599	560 (35%)	1,931	829 (43%)
Graminoid			534	61 (11%)	584	71 (12%)
Fern			102	5 (6%)	107	7 (6%)
State endemics in each floristic province						
Californian	970	137 (14%)	5,093	1,438 (28%)	6,266	2,226 (36%)
Desert	715	0 (0%)	2,437	171 (7%)	2,769	309 (11%)
State endemics in each biogeographic group						
North-temperate	373	15 (4%)	2,185	503 (23%)	2,643	767 (29%)
Subtropical semiarid	229	34 (15%)	1,849	794 (43%)	2,471	1,268 (51%)
Desert	171	0 (0%)	632	64 (10%)	742	112 (15%)

Top 5 families in numbers of Californian endemics

Asteraceae (341), Fabaceae (190), Polygonaceae (155), Boraginaceae (123), Polemoniaceae (116)

Top 5 genera in numbers of Californian endemics

Eriogonum (105), *Arctostaphylos* (79), *Astragalus* (70), *Lupinus* (57), *Clarkia* (50)

SOURCE: *The Jepson Manual: Vascular Plants of California*, 2nd ed. (Baldwin et al. 2012). Compression of the minimum taxonomic units treated in this database into full species, and determination of endemism for these full species, was done by Dylan Burge (unpublished). Life form data were obtained for 96% of species and taxa from CalFlora 2011. Ecoregions belonging to the California Floristic Province versus the Desert Floristic Province are from Baldwin et al. 2012. Biogeographic data are from Raven and Axelrod 1978; note that these three categories are not exhaustive.

TABLE 4   PLANT RICHNESS AND ENDEMISM IN THE WORLD'S MEDITERRANEAN
CLIMATE REGIONS

Region	Area (km²)	Number of Species	Endemism (%)
California	320,000	300	35
Chile	140,200	2,400	23
Cape	90,000	8,600	68
SW Australia	310,000	8,000	75
Mediterranean	2,300,000	25,000	50

SOURCE: Cowling et al. 1996.

Shrublands and the Great Basin Shrubsteppe ecoregions have more endemic plants (290 and 151, respectively) than any of California's 13 ecoregions, of which the richest is interior chaparral and woodland (150) (Ricketts et al. 1999). Other rich floras divided among multiple states are found in Appalachia, the southeastern coastal plain, the rock outcrops or "glades" within the eastern deciduous forest, and the "sky islands" of the southwestern deserts (Ricketts et al. 1999; Stein et al. 2000).

California is fourth among the world's five mediterranean climate zones in its plant richness and endemism (Table 4). Comparative evidence from these regions is discussed below.

There has never before been a comprehensive list of plants endemic to the California Floristic Province, including its Oregon and Baja California sectors. A new, preliminary list was generated based on the geographic information on plant distributions published in *The Jepson Manual,* 2nd ed. (Baldwin et al. 2012), and in the digital database of California plant taxa maintained by CalFlora (www.calflora.org). This information was supplemented by herbarium records for Oregon and Baja. For the work based on herbarium records, plant species were classified as CFP endemics if they were present in any of the spatial units belonging to the CFP portions of Baja and Oregon and absent from all spatial units immediately adjacent to these regions (using CFP boundaries approximating those in Figure 4, above). *The Jepson Manual* was used as the main authority for names of plants occurring in California. For plants occurring wholly outside California, names were taken from *Flora of Baja California* (Wiggins 1980), O'Brien et al. (2013), or the Oregon Flora Project. The list of species endemic to the CFP is undergoing continuing revision (see Appendix; Burge, Thorne, Harrison et al. unpublished), and the list for Baja California is especially likely to change with ongoing floristic study (see O'Brien et al. 2013).

TABLE 5   SUMMARY OF ENDEMISM IN THE CALIFORNIA FLORISTIC PROVINCE

Region	CFP Endemics	One-Region Endemics	Total Species
Baja California	338 (35%)	107 (11%)	969
Oregon	220 (15%)	14 (1%)	1,501
Cascade Ranges	297 (18%)	12 (1%)	1,636
Central Western	789 (40%)	175 (9%)	1,979
Great Valley	310 (29%)	25 (2%)	1,069
Northwest	724 (30%)	145 (6%)	2,422
Sierra Nevada	706 (27%)	219 (8%)	2,602
Southwest	735 (33%)	185 (8%)	2,214
ENTIRE CFP	1,931 (40%)		4,819

Using this preliminary list, 1,906 full species out of 4,571 total species were classified as CFP endemics, yielding an estimate of 41.7 percent endemism. For comparison, the last comprehensive estimate was 2,125 out of 4,452 species, or 47.7 percent endemism (Raven and Axelrod 1978). Baja adds 129 and Oregon 14 CFP endemic species to the Province that are not found in the state of California (Table 5).

GEOGRAPHY OF DIVERSITY AND ENDEMISM

In their classic analysis of Californian plant endemism, Stebbins and Major (1965) selected species or taxa using a priori judgments as to mode of endemism. They compiled 177 paleoendemics according to the criteria that the species either belong to monotypic genera or have their nearest relatives far away. Of these, the majority were north-temperate (Arcto-Tertiary), mostly either wetland species with relatives on the southeastern coastal plain or forest species related to the Appalachian or temperate Asian floras. The rest were considered subtropical semiarid (Madro-Tertiary) relicts having their closest relatives in the arid southwest but with more distant relatives in drier parts of Europe, South America, and/or North Africa. The greatest number of north-temperate relicts occurred in the North Coast region (including the Klamath-Siskiyou); of 68 such species, 12 were found only there. Paleoendemics were also abundant in the northern Sierra–Cascade (60) and Sierra (65), Central Coast (63), and Southern California (68) regions, as well as the Colorado (39), Mojave (33), and Inyo (45) Deserts, but not the Central Valley (13) or Great Basin (13) parts of the state. From north to south, an increasing proportion of paleoendemics were interpreted as being of southerly rather than Arcto-Tertiary origin.

Stebbins and Major (1965) then selected 70 species-rich Californian genera to represent neoendemics. The highest concentrations of these species occurred in the central Coast Ranges, the midelevation Sierran foothills, and the Southern California coast. Endemism was modest in the warm (Colorado and Mojave) deserts and low in the cold (Great Basin) deserts and Central Valley. The species considered "patroendemics," that is, relictual diploid ancestors of more widespread polyploid descendants, were most abundant in equable coastal climates. Those considered "apoendemics," or young polyploids recently derived from diploid ancestors, were concentrated in the harsher climates of the inner Coast Ranges. Ten hotspots of neoendemism were found in the Central Coast, all notable for their diversity of soils, topography, and climate. The richest hotspots were the Monterey region and the rugged Napa–Lake County border. Neoendemism, they concluded, was promoted by intermediate climates, where sharp gradients separate zones of moisture surplus and deficiency. In such settings, they suggested, plant species survived climatic fluctuations through localized shifts among varied soils, slopes, and elevations. As a result, genetic isolation and local adaptation alternate with secondary contact and hybridization, producing rapid speciation as well as rich gene pools for adaptation. Stebbins and Hrusa (1995) elaborated these ideas with respect to the southerly northern Coast Ranges, where they argued that the mild climate enhances differences in soil chemistry, texture, and topography, leading to rapid plant diversification during climatically driven fluctuations in the locations of boundaries between vegetation types.

A series of more recent analyses have used geographic databases to explore California's total and endemic plant diversity. Harrison and Grace (2007) found that Californian plant diversity at the regional scale is related to plant productivity, which is determined by rainfall. Species belonging to north-temperate genera that speciated extensively in the California Floristic Province, similarly to other north-temperate species, are increasingly diverse where rainfall and productivity are higher. Species in groups denoted by Raven and Axelrod (1978) as "strongly centered in the California Floristic Province," including many neoendemics, follow the same climatic patterns as other subtropical semiarid species; their diversity increases only weakly with rainfall and productivity. At least in these broad terms, the climatic niches of Californian endemics do not differ consistently from those of their nonendemic relatives.

Thorne, Viers et al. (2009) generated species distribution maps for 7,887 taxa (5,988 full species) in 228 geographic units across California. Endemics included 1,374 full species and 2,070 endemic taxa, most of which had small ranges (<10,000 km²). The Coast Ranges had the highest numbers of endemics, although the Sierra Nevada contained more total species. Concentrations of endemics with ranges greater than 10,000 square kilometers occur in the Channel Islands, the San Francisco Bay Area, western Riverside County, and the northern Coast Ranges. Endemism in the Klamath-Siskiyou region was noted to be underestimated because of the state boundary. The results agree with Stebbins and Major (1965) that the Coast Ranges are the richest center of endemism, but more endemism occurs in the Sierra and less in Southern California than those authors found by using a subset of endemics.

Kraft et al. (2010) analyzed the geographic distributions of 337 species, subspecies, and varieties considered neoendemic to the state for which molecular data was available to estimate their ages. The highest concentrations of neoendemics were found in the Central Coast (in particular, the San Francisco Bay Area), the southern part of the North Coast region, and the Sierra Nevadan foothills. Low neoendemism was found in all the nonmediterranean parts of the state. The neoendemics with the smallest ranges were found at low elevations in the Coast Ranges and at higher elevations in the Sierra and the Transverse Ranges. The youngest neondemics, however, occurred in the deserts. Neither the means nor the heterogeneity in physical variables (topography, climate) was a good predictor of endemic diversity.

Klamath-Siskiyous

The Klamath-Siskiyou region comprises much of northwestern California and southwestern Oregon, where a complex of mountain ranges (often designated the Siskiyous in the north and the Klamaths in the south) form the northernmost and rainiest part of the California Floristic Province. Botanists and biogeographers have long been aware of the Klamath-Siskiyou region as a center of narrow endemism, as well as species richness and diverse vegetation (Whittaker 1960, 1961). The majority of plants endemic to the region are confined to serpentine, limestone, gabbro, or other unusual soils (Whittaker 1960, 1961; Safford, Viers, and Harrison 2005). The traditional view of the region's diversity is that its rugged topography and near-maritime location have conferred a stable, well-buffered environment since the Tertiary, making it the most significant refuge for north-temperate taxa outside of Appalachia. It is also viewed as a meeting ground for plants of northerly, southerly, and Great Basin affinities (Whittaker 1960, 1961; Ricketts et al. 1999).

Smith and Sawyer (1988) estimated that the region from Roseburg, Oregon, in the north to Ukiah in the south (the Klamath-Siskiyou region plus the northern end of the Coast Ranges) harbors 3,500 plant taxa, including 281 endemics (173 full species). They noted that there are many examples of classic north-temperate taxa in the region (e.g. *Calypso, Darlingtonia, Berberis, Polystichum, Sequoia*), yet true paleoendemics found only there and having no nearby relatives are scarce: *Kalmiopsis leachiana, Picea breweri,* and *Quercus sadleriana* are among the few examples. There are also fewer endemic genera than might be expected in a climate refuge dating to the Tertiary. The genera with the most regional endemics for their sizes are *Lewisia, Sedum,* and *Sidalcea.* Other genera are rich in this region and elsewhere in the western United States, such as *Boechera* (*Arabis* sensu lato), *Penstemon,* and *Lupinus.* Smith and Sawyer (1988) concluded that the idea of this region as a museum of north-temperate relicts may be overstated. Nonetheless, because the Klamath-Siskiyou paleoendemics are so evolutionarily distinctive, they add disproportionately to perceptions of an ancient flora.

Sierra Nevada

Higher elevations of the Sierra Nevada have been described as geologically young and therefore poor in endemics (Stebbins and Major 1965; Raven and Axelrod 1978). However, Thorne, Viers et al. (2009) note that this conclusion may have followed from the fact that Stebbins and Major (1965) selected primarily lowland genera to represent Californian neoendemics. Underestimation of Sierran endemism also arises from the region's position along a state boundary. Chabot and Billings (1972) found that nearly one-sixth of the alpine flora near Piute Pass were endemic and concluded there had been substantial in situ evolution of lineages that colonized from lowland sources into the young and isolated alpine environment. Sierran endemics come from western sources (e.g., *Calyptridium,* which was found near the coast in historic times) as well as Great Basin sources (e.g., *Eriogonum, Cryptantha,* and *Ivesia*). Compared to other North American mountain ranges, the Sierra has a milder climate that may have facilitated colonization from the lowlands.

Shevock (1996) found that the Sierra as a whole harbored 405 endemic vascular plants, or 15 percent of its flora. There is a general increase from north to south in total species, Sierran endemics, and river basin endemics, with the Kern and Kings River basins being outstanding for richness and endemism. The genera with the most rare and/or endemic taxa are *Erigonum, Astragalus, Mimulus,* and *Clarkia.* Endemic trees are Sierra foxtail pine, Piute cypress, and giant sequoia (*Pinus balfouriana* ssp. *austrina, Hesperocyparis nevadensis, Sequoiadendron giganteum*), all found in the southern part of the range. Rare and endemic plants are found in many vegetation types, including yellow pine (211 species), foothill woodland (139), subalpine forest (124), meadow (116), and chaparral (90), and within these, about 12 percent are restricted to outcrops of carbonate rocks, serpentine, granite, or basalt.

Channel Islands

The California Channel Islands harbor at least 62 full species of plants restricted to one or more of the islands (listed in Schoenherr et al. 1999). A list that includes named subspecific taxa and also includes Cedros and Guadalupe Islands—Mexican islands with desertlike levels of precipitation but with thorn scrub and chaparral vegetation owing to their cool maritime temperatures—finds 174 endemics out of 920 taxa on the islands (Oberbauer 2002). The southern four Channel Islands are the most isolated, and according to Raven (1965), these support 42 endemic full species, including 6 restricted to San Clemente, one to Catalina, 2 to San Nicolas, one to Santa Barbara, 16 to several southern islands, and 11 to both northern and southern islands. The northern four islands have 23 endemic full species, including 5 restricted to Santa Cruz, 7 to several northern islands, and (again) 11 shared with the southern group of islands. The only island endemic genera still recognized are *Lyonothamnus* (Rosaceae), on both Catalina and Santa Cruz; and *Munzothamnus* (Asteraceae), on San Clemente. (Two endemic genera are known from Guadalupe: *Hesperelaea* [presumed extinct], Oleaceae; and *Baeriopsis,* Asteraceae.) Many island endemics belong to classic Californian neoendemic genera such as *Cryptantha, Gilia, Malacothrix,* and *Phacelia. Dudleya* has 5 endemic species and 4 endemic subspecies on the islands.

The overall endemism of the Channel Islands has been described as low, in fact not strikingly higher than the adjacent mainland (Raven 1967). Some islands were thought to have been connected to the mainland in the Pleistocene, although it is now believed they were as close as 6 kilometers but not connected. The four smallest islands were submerged until the mid-Pliocene. Only San Clemente and Guadalupe are large, old, and isolated enough to have substantial endemism. Raven (1967), R. F. Thorne (1969), and subsequent authors concluded that the island flora has been more strongly shaped by relictual survival in the cool maritime climate than by insular evolution per se. Endemic herbs on the island are often replaced by a more drought-adapted congener on the nearby mainland and/or have close relatives much farther north along the coast. Other mesic northerly taxa are found on the islands only as fossils, such as species of *Pseudotsuga, Garrya, Ceanothus,* and *Hesperocyparis* (*Cupressus* sensu lato).

Recent evidence suggests that island evolution may have been underestimated. The California Islands' perennial tarweeds (*Deinandra* = *Hemizonia* sensu lato), woody species in an otherwise herbaceous clade, include the endemics *D. clementina, D. frutescens, D. greeneana,* and *D. palmeri.* These were once considered primitive forms that survived on the islands (Oberbauer 2002), but new analyses show that they represent insular adaptive radiation and derived growth forms (Baldwin 2007). Likewise, on the basis of Nevadan fossils, the endemic ironwood tree, *Lyonothamnus floribundus,* was considered a classic paleoendemic that ranged widely on the mainland in the Pliocene and became restricted to the islands as the climate became drier, but the extant species and its two subspecies are now believed to have evolved on the islands (Erwin and Schorn 2000).

A number of species are endemic not to the Channel Islands per se but to the coastal fog belt on the islands and/or nearby mainland (e.g., *Pinus torreyana, Salvia brandegeei, Leptosyne* (*Coreopsis*) *gigantea, Eriogonum grande, Eriodictyon traskiae, Hemizonia greeneana, Prunus ilicifolia* ssp. *lyonii, Senecio lyonii*). Some of these have islandlike traits such as gigantism or woodiness (in otherwise small herbaceous lineages) and would have been considered classic cases of island evolution if their few mainland populations had not been discovered (Thorne 1969). The fog belt is an important refuge for species intolerant of summer drought. Studies of the endemic Bishop pine (*P. muricata*) show that fog drip plus cloud shading can reduce annual drought stress by more than 50 percent (Fischer et al. 2009). The foggy climate causes nearly year-round flowering in many Channel Island species whose mainland relatives are strongly seasonal, for example, *Dendromecon harfordii* and *Galvezia speciosa* (Thorne 1969).

EDAPHIC ENDEMISM

Specialization to particular edaphic (bedrock or soil) substrates is probably higher in California's flora than that of almost any floristic province in the

TABLE 6   NUMBERS OF CALIFORNIA SPECIAL-
STATUS PLANT TAXA ASSOCIATED WITH
UNUSUAL SOIL OR BEDROCK SUBSTRATES

Soil/Substrate	Number
Serpentine	298
Granite	126
Clay	113
Carbonate	111
Volcanic	99
Alkaline	82
Vernal pool	61
Gabbro	20
Sandstone	19
Shale	10
Gypsum	1

SOURCE: CDFW 2003; vernal pool estimate from CNPS
2011.

world. One way to quantify this is with the list of special-status taxa compiled by the California Native Plant Society, which includes rare, threatened, and endangered species, subspecies, and varieties (CNPS 2001). The
current list of 2,058 special-status taxa exceeds that of any other state and
accounts for 30 percent of the flora. According to the CNPS (2001), as
cited by the CDFW (2003), 879, or 43 percent, of California's special-status plant taxa are edaphic specialists (Table 6). If these are representative of
the flora as a whole, then 43 percent—or almost 3,000 species, subspecies,
and varieties—of the state's plants are edaphic specialists. More likely,
however, edaphic specialists have stronger tendencies to achieve "special
status" than edaphic generalists. But even if there are no edaphic specialists
other than the 879 taxa in Table 5, these are an impressive 13 percent
(43% × 30%) of the total flora. Serpentine, limestone, gabbro, vernal
pools, and other unusual substrates supporting endemic plants in California are described in rich detail by Kruckeberg (2006).

Serpentine (Figure 11) is the most chemically and floristically distinctive, broadly distributed, and frequently studied special edaphic substrate
in California. As used by ecologists, "serpentine" refers to the ultramafic
(high iron and magnesium) rocks peridotite and serpentinite and to the
soils formed from the weathering of these rocks. Peridotite constitutes
much of the earth's mantle and oceanic crust, and serpentinite is derived
from peridotite by hydration. Both tend to occur in long, discontinuous
strings on continental land surfaces that mark the locations of present or

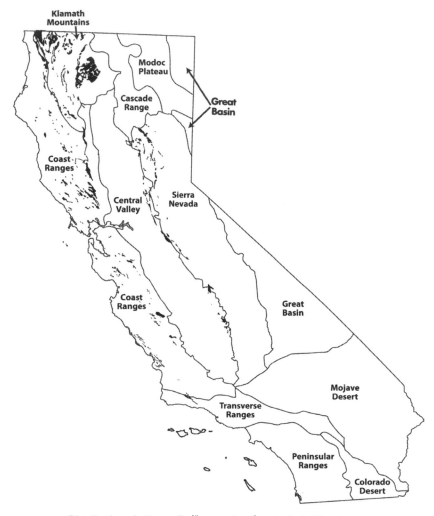

FIGURE 11. Distribution of ultramafic ("serpentine") rocks in California.

former subduction zones (Kruckeberg 1984, 2006; Harden 2004; Alexander et al. 2006; Harrison and Rajakaruna 2011). Using a comprehensive survey of literature and herbarium data, Safford et al. (2005) enumerated 164 "strict endemics" and 82 "broad endemics" to serpentine in California, for a total of 246 taxa that include 176 full species. An additional 150 taxa are "strong indicators" and 123 are "weak indicators" of serpentine. These authors estimated that 12.5 percent of plant species endemic to California are either strictly or broadly endemic to serpentine, even though serpentine is less than 2 percent of the state's area. Serpentine endemics are particularly rich in the Klamath-Siskiyous and northern Coast Ranges (Safford et al. 2005).

Plants may be restricted to soils such as serpentine by competition from other plants that grow faster on more productive soils but that are less tolerant of infertile soils. Kruckeberg (1954) showed that a serpentine endemic jewelflower (*Streptanthus*) grew well on fertile nonserpentine soils, and on serpentine soils with fertilizer added, but only as long as it was free from faster-growing competitors. Later work generally, though not entirely, supports the premises that serpentine endemics do not physiologically require serpentine (Brady et al. 2005), that competition is reduced on serpentine (Elmendorf and Moore 2007), and that plants growing on serpentine have functional traits conferring stress tolerance at the expense of the capacity for rapid growth, for example, low specific leaf area, low foliar nitrogen content, and short stature (Spasojevic et al. 2012).

Raven and Axelrod (1978) used the idea of serpentine (and other infertile substrates) as refuges from competition in their broader biogeographic theory of the Californian flora. They argued that serpentine endemics with relatively large ranges, mainly shrubs such as *Quercus durata* and *Hesperocyparis* (*Cupressus*) *sargentii*, invaded serpentine as it became available through mountain uplift in the late Pliocene. These species then became extinct on nonserpentine as the summer drought increased and persist today on serpentine where the dominant vegetation is not able to compete well. This argument rests on an assertion that serpentine environments are unusually moist to wet, allowing paleoendemics to find refuge in a drying climate; this is difficult to reconcile with the perception by field biologists that serpentine environments are hot and dry, with sparse cover and vegetation that appears drought-adapted (e.g., Whittaker 1960; Kruckeberg 1984). In contrast to the widely distributed serpentine paleoendemics, Raven and Axelrod (1978) noted that the majority of serpentine endemics (e.g., in *Navarretia, Streptanthus, Allium, Hesperolinon*, and other genera) have narrow distributions and probably arose by ecological speciation in late Pleistocene to recent time.

Recent phylogenetic analyses shed new light on the origin of Californian serpentine endemics. Examining 23 genera, Anacker et al. (2011) found that both tolerance and endemism to serpentine often arose multiple times within a lineage but that endemism tended to be an evolutionary dead end causing reduced diversification. Endemics appear equally likely to arise in two ways: from ancestors intolerant of serpentine, where speciation accompanied the evolution of serpentine

TABLE 7   PLANT SPECIES ENDEMIC TO CALIFORNIA
VERNAL POOLS

Family	Species
Apiaceae	*Eryngium constancei*
	Eryngium stenosepalum
Asteraceae	*Blennosperma bakeri*
	Lasthenia burkei
	Lasthenia conjugens
	Lasthenia ferrisiae
	Lasthenia fremontii
Boraginaceae	*Plagiobothrys acanthocarpus*
Campanulaceae	*Downingia bella*
Cuscutaceae	*Cuscuta howelliana*
Euphorbiaceae	*Chamaesyce hooveri*
Lamiaceae	*Pogogyne abramsii*
	Pogogyne nudiuscula
Limnanthaceae	*Limnanthes vinculans*
Phrymaceae	*Mimulus tricolor*
Plantaginaceae	*Gratiola heterosepala*
Polemoniaceae	*Navarretia fossalis*
	Navarretia leucocephalus
	Navarretia myersii
	Navarretia nigelliformis
Poaceae	*Agrostis hendersonii*
	Agrostis tandilensis
	Neostapfia colusana
	Orcuttia californica
	Orcuttia inaequalis
	Orcuttia pilosa
	Orcuttia tenuis
	Orcuttia viscida
	Tuctoria greenei
	Tuctoria mucronata

SOURCE: Keeler-Wolf et al. 1998.

tolerance, or from serpentine-tolerant ancestors, where restriction to serpentine entailed loss of ability to grow on other soils, perhaps due to changes in the competitive environment. In the first type of transition (nontolerator-to-endemic), although not in the second (tolerator-to-endemic), endemics are found in cooler, wetter, and less seasonal environments than their nonendemic relatives (Anacker and Harrison 2012). This suggests a different link between climate and endemism than Raven and Axelrod (1978) envisioned: benign, wetter climates may promote the survival of small populations with novel adaptations such as serpentine tolerance (Anacker and Harrison 2012).

Evidence for the evolution of serpentine endemics, including the roles of fitness tradeoffs and pleiotropy, phenological divergence, catastrophic selection, and spatial isolation, has recently been reviewed by Kay et al. (2011). Some examples are discussed below (see especially Tarweeds). Serpentine may provide some of the best cases of ecological speciation in which there is relatively little spatial isolation between progenitor and derivative species and selection against hybrids is strong enough to overcome the homogenizing effects of gene flow (Kay et al. 2011).

The dwarf wild flax genus *Hesperolinon* consists of 9 serpentine endemics and 4 serpentine tolerators. Phylogenetic analyses suggest the repeated evolution of narrow endemism from generalist ancestors (Springer 2009). The genus has probably evolved recently and rapidly, as evidenced by the many unresolved relationships among taxa. *Hesperolinon* species growing on the most stressful (low-calcium) serpentine soils had lower incidence and severity of flax rust disease, suggesting that serpentine may act as a refuge from pathogens as well as competitors (Springer 2009).

Vernal pools have over 100 plant species and subspecific taxa associated with them, but just 32 full species are considered obligate vernal pool endemics (Table 7). The evolutionary and ecological processes giving rise to vernal pool endemism have been studied in a few clades, discussed below (see especially *Limnanthes*).

NEW EVIDENCE ON THE EVOLUTION OF NEOENDEMICS

Molecular techniques have made it possible to determine the patterns and timing of diversification much more explicitly than in the past, and to test for congruence between genetic divergence within lineages, called breaks, and geologic and climatic processes. As these studies have proliferated, it has even become possible to seek broad patterns shared by groups of plants and animals within a region. These multilineage analyses are examined first below, followed by studies of individual clades of Californian plants.

Analyses across Multiple Lineages

Calsbeek et al. (2003) analyzed molecular data from published studies of 55 Californian plant and animal lineages. In the six Californian plant genera studied (*Calycadenia, Linanthus* sensu lato [including

Leptosiphon], *Gilia, Ceanothus, Lithophragma,* and *Calochortus*), the average age of divergences was younger than in animals, around one million as opposed to 5 to 2 million years. In contrast to animals, the plants showed little congruence in the location and timing of genetic breaks, either with one another or with the emergence of mountain and water barriers. For plants, these results suggest that relatively recent climatic oscillations may have been more important in diversification than large-scale geologic barriers. However, in four genera (*Calycadenia, Linanthus, Gilia, Ceanothus*), the initial onset of diversification was older than for most animals (7–17 million years), suggesting that the gradually drying Miocene climate set the stage for modern diversification.

In analyses of the geographic hotspots of neoendemism, described earlier in this chapter, Kraft et al. (2010) used molecular data to estimate the ages of plant taxa considered neoendemic to the state. The average estimated ages were 2.8 to 5.0 million years for 236 neoendemic herbs and 2.5 to 5.4 million years for 101 neoendemic shrubs and trees. These results suggest that much of the speciation leading to neoendemics took place in the premediterranean climate of the Pliocene, when aridity was increasing and most but not all precipitation fell in winter.

Using molecular phylogenetic evidence for 16 large plant clades, Lancaster and Kay (2013) revisited some of Raven and Axelrod's (1978) major conclusions about the origins and diversification of the Californian flora. The focal clades contained 1687 species, with each clade found both inside (444 species) and outside (1,243 species) California. Across all clades combined, extinction rates were lower inside than outside California, but, surprisingly, speciation rates were not higher in California. Within California, the north-temperate lineages manifested lower speciation rates than Madrean lineages, but extinction rates in California did not differ among lineages. The mean ages of residency in California were estimated as 24, 27, and 6 million years for north-temperate, subtropical semiarid, and desert lineages, respectively. These authors concluded that, in contrast to the view of Raven and Axelrod (1978) and many other authors, high endemism in California is less the product of rapid speciation under a mediterranean climate and more that of lower extinction rates across all clades caused by the historic stability of the climate. In this view, high diversity substantially preceded the modern climate. The results did support Raven and Axelrod's (1978) assertions that the north-temperate and subtropical semiarid lineages are relatively old and that the desert lineages are recent arrivals. However, all clades and not just north-temperate lineages experienced reduced extinction in California.

Lancaster and Kay (2013) note that certain plant clades do show elevated rates of speciation in California, for example, the clades containing *Mimulus* and some Asteraceae. In California and in other mediterranean regions, diversification and endemism tend to be highly concentrated within a few clades, some of which may be entirely endemic, so a complete sampling of these clades would be valuable. Also, the results do not rule out that high diversity is caused by the high representation in California of certain clades, such as annual plant genera, that have high speciation rates both inside and outside the state. Nevertheless, this study is among the first comparison of diversification rates inside and outside any center of endemism and provokes reconsideration of the classic story of the Californian flora.

The finding by Lancaster and Kay (2013) that speciation rates in California were lower for north-temperate taxa is consistent with Raven and Axelrod (1978). However, the latter authors noted that around 50 north-temperate genera have speciated extensively in California (e.g., *Allium, Calochortus, Iris, Lomatium, Senecio*) (Raven and Axelrod 1978: 16). These genera contain 645 full species, of which 253 (34%) are strictly (neo)endemic to California. Since there are a total of 503 Californian endemics among the 2,185 north-temperate species, the north-temperate paleoendemics must comprise only around 250 species, or 503 minus 253. In contrast, of the 1,849 subtropical semiarid species, 794 (43%) are endemic to California. We can conclude from this that there are some north-temperate lineages that are as speciation-prone as the subtropical semiarid lineages, and indeed, that neoendemism is as common as paleoendemism even in the north-temperate lineages—contrary to the conventional view of these lineages. The survival of the mesic Tertiary flora may have contributed substantially to the evolution of new species in California.

Analyses of Individual Lineages

Tarweeds (Madiinae)

The Asteraceous clade Madiinae is unusually well studied and demonstrates multiple patterns of diversification (Baldwin 2006). It includes the tarweeds, which are mostly Californian annuals with sticky glandular foliage, and the Hawaiian silverswords, large and morphologically diverse perennials that arose from a tarweed ancestor and underwent a spectacular island radiation. The tarweed genus *Layia* includes

FIGURE 12. Evolution of an endemic tarweed: (a) the serpentine endemic *Layia discoidea* (photo by Aaron Schusteff); (b) its ancestor, *L. glandulosa* (photo by Paul V. A. Fine).

FIGURE 12 *continued.* (c) A possible example of hybridization, in which *L. discoidea* displays the ray flowers of *L. glandulosa* (photo by Ryan O'Dell, Bureau of Land Management, Hollister Field Office). Photos a, b, and c are from the Clear Creek (New Idria) region of San Benito County, California.

14 species, of which 11 are endemic to mediterranean regions of California. Six *Layia* species that have seven chromosome pairs form a clade in which genetic differentiation, morphological differences, and interfertility are all highly congruent, consistent with gradual allopatric divergence (geographic speciation). Sympatric sister taxa are older than allopatric ones, supporting the idea that sympatry is secondary. Relatively concordant molecular, chromosomal, and morphological differentiation is also found in the 15 species of the tarweed genus *Calycadenia* (Baldwin 1993), except in one clade with striking differences in relative rates of morphological and chromosomal evolution (*C. multiglandulosa*, *C. pauciflora*, and relatives; see Baldwin 2003).

The clade of eight *Layia* species having eight chromosome pairs, in contrast, shows little congruence between phylogeny, morphology, and interfertility. Morphologically and ecologically novel species are embedded within widespread ancestral species, such as the serpentine endemic *Layia discoidea* (Figure 12a) within *L. glandulosa* (Figure 12b) and dune-inhabiting *L. carnosa* within *L. gaillardiodes*. Interfertility may be either high, as in the first pair, or low, as in the second. Likewise, the widespread tarweed *Holocarpha virgata* contains many cryptic, intersterile lineages, plus one morphologically novel descendant species with which it is interfertile, *H. macradenia.* While *H. virgata* is broadly distributed in interior grasslands, *H. macradenia* is confined to central coastal terraces, where it may have invaded and evolved during Holocene warming. These are some of the best-studied cases of ecological speciation in peripheral isolates. Whereas the classic model of catastrophic speciation called for major genetic reorganization of the newly evolved species, these tarweed examples demonstrate a stronger role for ecological selection (Baldwin 2006). Evolution has been both more rapid and more heterogeneous in tarweeds than had been envisioned by classic authors (e.g., Clausen 1951).

Californian tarweeds crossed the ocean to give rise to the Hawaiian endemic silversword lineage approximately 5 million years ago. Like many young and rapidly evolving woody lineages, they are interfertile, which facilitates hybrid speciation followed by rapid adaptation to new niches. Their closest mainland relatives include the genus *Madia,* which consists of ecologically similar and intersterile annuals. The contrast between the highly similar Californian and highly divergent Hawaiian Madiinae demonstrates that ecological opportunity, or vacant niche space, was far lower in California than on true oceanic islands, producing a much lesser degree of adaptive radiation (Baldwin 2006).

In perhaps the best-studied case of progenitor-derivative speciation in a Californian endemic, recent molecular work has revised and extended older conclusions about how serpentine-endemic *Layia discoidea* (Figure 12a) arose from *L. glandulosa* (Figure 12b) on a serpentine outcrop in the central Coast Ranges (Baldwin 2005). They are strongly differentiated, with *L. discoidea* lacking rays and *L. glandulosa* having either white or yellow rays. Intermediate forms suggestive of hybridization (Figure 12c) are rare. A yellow-flowered population of *L. glandulosa* appears to have given rise to *L. discoidea* less than one million years ago. The two lineages appear to be completely genetically isolated, even though they are fully interfertile and occur within a few kilometers of one another (Figure 13), potentially illustrating the role of strong natural selection against hybrids.

Clarkia

The genus *Clarkia* includes 42 species, of which 39 are found in the California Floristic Province and 36 are endemic to it; Raven and Axelrod

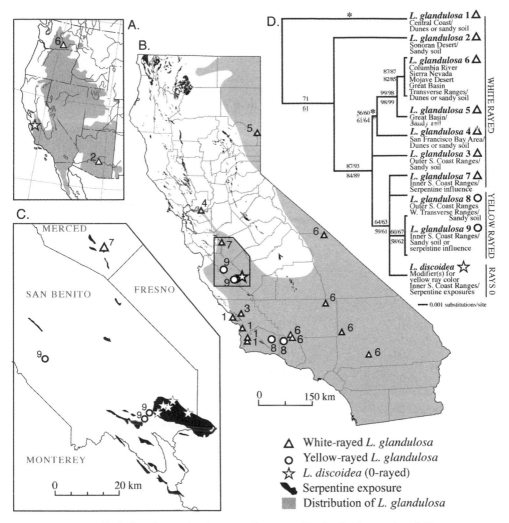

FIGURE 13. Evolution of an endemic tarweed: geographic distributions and evolutionary relationships of the serpentine endemic *Layia discoidea* and its ancestor, *L. glandulosa*. (Source: Reproduced by permission from Baldwin, B.G., 2005. Origin of the serpentine-endemic herb *Layia discoidea* from the widespread *L. glandulosa* (Compositae). Evolution 59: 2473–2479].

(1978) considered it a classic example of a group that radiated in the Madrean vegetation of western North America. Although these species vary in their degree of morphological distinctness, they usually differ in chromosomal structure and are interrelate. Classic studies reviewed by Gottlieb (2003) used *Clarkia* as an example of progenitor-descendant speciation. The best-studied species are *C. lingulata,* narrowly distributed in the Merced River canyon, and its presumed progenitor, *C. biloba*. These species inspired the idea of catastrophic speciation, whereby chance chromosomal rearrangements in small and disjunct populations produced new derivative species no better adapted to their environment than their progenitor. Newer studies have confirmed the progenitor-descendant relationship of these species and the absence of major novel adaptations in *C. lingulata*. However, the cause of the chromosomal rearrangement and the relative roles of selection, selfing, and spatial isolation in the origin and maintenance of the descendant remain unclear, in comparison to the clear role for natural selection in *L. discoidea* (Gottlieb 2003).

Mimulus

The monkeyflower genus, *Mimulus,* has around 120 to 200 species worldwide, with about 75 percent belonging to a western North American clade and the rest to a western Australian clade (Beardsley et al. 2004; Wu et al. 2007). They are extremely diverse in their habitats, which include desert, mediterranean semiaquatic, alpine, and "extreme" soil environments. They are also diverse in morphology, including flower color and shape, phenology, growth form, and breeding systems. Because they are also crossable and transplantable and have well-developed markers, the genus has become a model system for studies of the genetics of speciation, inbreeding, mating system evolution, and tolerance to extreme environments.

Studies of North American *Mimulus* have demonstrated the importance of chromosomal changes in speciation (Beardsley et al. 2004; Wu et al. 2007). The ancestral number ($n = 8$) was modified by at least 13 polyploid and 15 aneuploid (mismatched chromosome) events, estimated to have caused 11.5 percent (13/113) and 13.3 percent (15/133) of speciation events, respectively. Interestingly, most of these events occur near the tips of the phylogeny, suggesting that they seldom give rise to long-lasting and diverse lineages. Some of these events may have happened through hybridization, the rest through within-species genomic duplications; hybrid origin is strongly supported for only one species (*M. evanescens;* Beardsley et al. 2004). Multiple independent origins of polyploidy in *Mimulus* provide the opportunity to see whether gene rearrangement after duplications follows particular rules, how duplicate gene loss might lead to reproductive isolation within polyploid lineages, and how polyploidy accelerates evolution by increasing variability and creating transgressive phenotypes (Wu et al. 2007).

Reflecting its evolutionary dynamism, *Mimulus* includes more rare taxa with unresolved taxonomic status than any other Californian genus (29) (Skinner et al. 1995). Taxonomy is especially problematic in the woody section of *Mimulus* (*Diplacus*), which has from 2 to 13 named taxa that are not monophyletic and appear to interbreed freely in disturbed or intermediate habitats, suggesting recent divergence and secondary contact. Another section (*Eunanus*) shows evidence that morphological change has been accelerated relative to genetic change in the California lineage (Beardsley et al. 2004).

Ecological speciation has been studied in the yellow-flowered *M. guttatus* group, which contains much of the ecological diversity of the genus. *M. guttatus* is widespread, outcrossing, and genetically diverse within and among populations. Its close relatives have narrower ranges, less genetic diversity, and more selfing; many are habitat specialists on seashores, alpine zones, deserts, basalt cliffs, copper mine tailings, sand dunes, serpentine barrens, grasslands, alpine meadows, edges of geysers, or peat bogs (Wu et al. 2007). Some populations of *M. guttatus* are locally adapted to copper mine tailings or serpentine soils, yet there are also two apparent derivative species that specialize on these environments: *M. cupriphilus* on copper and *M. nudatus* on serpentine. Because the derivative species are highly selfing, it is difficult to explore their relationships to *M. guttatus* using crossing experiments. However, studies of copper-tolerant and copper-intolerant *M. guttatus* suggest the importance of allelic incompatibility in causing hybrid lethality. The incompatible alleles may be nuclear or cytoplasmic, and are not found in all the tolerant populations; restorer genes can also evolve, leading to complex patterns of hybrid compatibility (Wu et al. 2007). Other evidence suggests that in some cases adaptation to variable levels of drought stress is correlated with divergence in life history, mating system, flowering time, and/or morphology, facilitating speciation (Wu et al. 2007). For example, Lowry et al. (2008) found that selection on flowering time was crucial to differentiating large coastal *M. guttatus grandis* from smaller inland subspecies, and Hall et al. (2006, 2010) identified some of the genes responsible for local adaptation and reproductive isolation.

Limnanthes

The meadowfoam genus *Limnanthes* and its family, Limnanthaceae, are nearly endemic to the California Floristic Province, except for one *Limnanthes* species found in the Pacific Northwest and one monotypic genus (*Floerkea*) found in California and in the northern and eastern United States. The nine currently recognized Californian *Limnanthes* all occur in vernal pools or wet meadows. The genus was studied by Mason (1952) and Ornduff (1971) using plant morphology, crossing experiments, and other evidence, and by later authors using flavonoid biochemistry, allozymes, and nuclear and chloroplast DNA (Meyers et al. 2010). Consistent support has been found for the subdivision of *Limnanthes* into the two sections originally proposed by Mason (1952), but there is little resolution within these sections. Successive studies have disagreed on the boundaries and patterns of relatedness among species and subspecies and have reached few clear conclusions as to the overall modes of evolution in the genus. The most recent molecular evidence suggests that there are only four clear monophyletic species of *Limnanthes* (Meyers et al. 2010). Genetic differentiation among conspecific populations is higher in *Limnanthes* than most herbs, probably because of its fragmented distribution, and gradual allopatric divergence seems likely but unproven (McNeill and Jain 1983).

Leptosiphon (Linanthus sensu lato)

The genus *Linanthus,* as it was then called, was noted by Raven and Axelrod (1978) as a classic subtropical semiarid neoendemic genus; 32 of its 37 then-extant species were confined to the California Floristic Province, and except for one in Chile the rest were elsewhere in the western United States. They believed this genus and others in the Polemoniaceae originated in subtropical woodland, perhaps 20 million years ago, began diversifying as the summer-dry climate began to develop between 13 million and 15 million years ago, and speciated rapidly when the mediterranean climate took hold. Bell and Patterson (2000) examined the molecular phylogenetics of *Linanthus,* which by then contained 44 recognized species, of which 36 are found in the California Floristic Province. Given the absence of a fossil record, these authors used the onset of the winter-rainfall climate 15 million years ago to constrain the date of divergence of *Linanthus,* plus the closely related *Phlox* and *Leptosiphon,* from their sister clade, *Polemonium.* The results showed that *Linanthus* actually consists of two lineages, one more closely related to *Gilia* and the other to *Phlox,* and subsequent revision placed most of the Californian species into *Leptosiphon.* The data remain consistent, however, with Raven and Axelrod's belief that the entire group containing *Linanthus* began to diversify rapidly in the

Pliocene. *Linanthus* (broad sense) has been the subject of evolutionary studies on plant-pollinator interactions (Schmitt 1983), breeding systems (Goodwillie 1999), flower color (Schemske and Bierzychudek 2001), and serpentine endemism (Kay et al. 2011).

Ceanothus

With 38 of its 53 species endemic to the California Floristic Province, the California-lilac (*Ceanothus*) is perhaps more emblematic of California than any other chaparral shrub, with the possible exception of manzanita (*Arctostaphylos*). Its two subgenera, *Euceanothus* (called *Ceanothus* by some authors, but *Euceanothus* is used here for clarity) and *Cerastes,* differ markedly; *Euceanothus* is typical for the family Rhamnaceae with its broad and thin leaves, while *Cerastes,* with small, thick, often spiny leaves having sunken stomatal crypts, appears more drought-adapted. Although the two subgenera are intersterile, species within each subgenus are interfertile and prone to forming hybrid swarms, and hybrid origins have long been suspected for several species in *Cerastes.*

Molecular phylogenetic analysis of *Ceanothus* supports the monophyly of the two subgenera, but other relationships remain unresolvable (Hardig et al. 2000; Burge et al. 2011). Patterns in nuclear and organellar DNA are often noncongruent, which can be an indication of hybridization between adjacent species (e.g., *C. verrucosus* and *C. crassifolius*), although it is more likely to be caused by incomplete lineage sorting when the species in question are widely separated (e.g., *C. hearstiorum* and *C. lemmoni*). Molecular data do not support the old idea that the subgenus *Cerastes* arose from within the subgenus *Euceanothus*. The numerous, narrowly distributed species of the subgenus *Cerastes* probably arose from within a progenitor species that is now paraphyletic, contributing to the lack of resolution at the species level (Hardig et al. 2000). Divergence times estimated from a combination of molecular evidence and fossil calibration place the split between subgenera *Euceanothus* and *Cerastes* at approximately 13 million years ago, during the cool, dry Miocene (Burge et al. 2011). Within each subgenus, the onset of the diversification leading to modern taxa was estimated at 6 million years ago, consistent with most diversification having occurred as the mediterranean climate developed and the Coast Ranges uplifted (Figure 14; Burge et al. 2011).

Using the phylogenetic tree provided by Hardig et al. (2000), Ackerly et al. (2006) found that *Ceanothus* species coexisting in a single

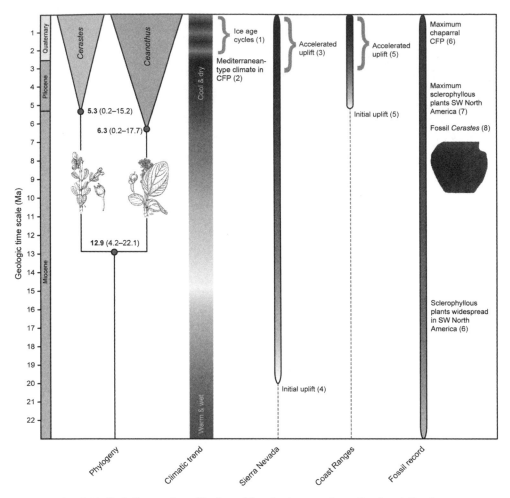

FIGURE 14. Evolutionary diversification of the shrub genus *Ceanothus* in relation to geologic and climatic events. (Source: Burge et al. 2011)

community are nonrandomly likely to be drawn from the two different subgenera and to differ in leaf traits that evolved early in the history of the genus, although they resemble each other in their climatic distributions, which evolved later. Thus, the early divergence of the two subgenera with their different leaf strategies may have enabled both to radiate in California while coexisting.

Wells (1969) argued that the massive speciation of *Arctostaphylos* and *Ceanothus* in California is related to the loss of the ancestral ability to survive fire by resprouting. About four-fifths of the species in these two very species-rich (>50 species) genera reproduce after fire by seed, whereas other prominent chaparral shrub genera are resprouters and have only 1 to 7 species each. Obligate postfire seeding shortens generation time, increasing both the frequency and intensity of natural selection on seedlings and therefore the potential rate of evolution. Whether such evolution can be considered an adaptive radiation is questionable; Wells (1969) points out that generalist resprouters such as *Adenostoma fasciculatum* (chamise) are at least equal in geographic distribution and total numbers to *Arctostaphylos* and *Ceanothus*.

NEW EVIDENCE ON PALEOENDEMISM

In the past few decades, the reconstruction of California's floristic history has benefited greatly from several developments (reviewed by Millar 1996): the availability of fossil pollen from marine sediment cores, radiometric dating of fossils, and greatly improved global and regional climate reconstruction from isotope ratios in ice and sediment cores. Evidence for the oscillation of dominant vegetation during Pleistocene glacial-interglacial cycles comes from pollen and isotopes in sediment cores, especially a 160,000-year record from the Santa Barbara Basin, and also from Clear Lake, Owens Lake, and San Miguel Island (Millar 1999). Combined with the steady discovery of new fossils and other evidence (e.g., molecular analyses of population relationships), this has led to revisions in the understanding of the paleoendemic component of the flora.

Conifers endemic to the coastal fog belt include redwoods (*Sequoia sempervirens*), Torrey pines (*Pinus torreyana*), and the closed-cone Monterey pine (*P. radiata*) and Bishop pine (*P. muricata*). Redwoods are an archetypal north-temperate species with a fossil record dating to the Cretaceous (Sawyer et al. 2000). Widespread across the mid-latitude continental interior in the Eocene, when its relative *Sequoiadendron* (giant sequoia) may have inhabited higher latitudes, the fossil species *Sequoia affinis* (similar or identical to *sempervirens*) contracted and shifted to the south and west throughout the Tertiary and attained its present coastal distribution in the Pleistocene. From fossil pollen in offshore marine sediments, more details of its recent history are now known. It extended south to Santa Barbara in the Pleistocene (as did Monterey pine). It declined during glacial periods when the coast was dominated by spruce, fir, and Douglas fir and reached maximum abundance in the mid-Holocene when a warm climate and strong upwelling produced a more extensive coastal fog zone than exists today (Sawyer et al. 2000).

The closed-cone pines immigrated from Mexico in the Oligocene and were once believed to have to have formed near-continuous stands along the coast during the Pleistocene until they become fragmented into today's disjunct stands by Holocene warming. Molecular evidence still supports the Mexican origins, age, and close relatedness of these species. However, newer evidence reveals a dynamic Pleistocene history in which *Pinus radiata* was always patchily distributed, reaching its highest abundance in times of intermediate temperature and becoming replaced by junipers during colder periods and by oaks and Asteraceae in warmer periods of the Pleistocene (Millar 1999).

CALIFORNIAN PLANT ENDEMISM IN A WORLD CONTEXT

All five mediterranean climate regions of the world are exceptional for their plant endemism, and in fact are the major outliers to the global relationships between plant diversity and area, latitude, and productivity. Among the five regions, the richest are the Cape Floristic Region and southwestern Australia, followed by the Mediterranean Basin and California (roughly tied) and central Chile (Cowling et al. 1996). Within each region, the geographic and taxonomic distribution of diversity and endemism also provide clues to the processes underlying diversification.

In all five regions, the climate is believed to have become fully mediterranean only within the past 3 million (or fewer) years and to have undergone recurrent ice ages in the past 2 million years, with mean annual temperature fluctuations of 2° to 7°C. Endemism is concentrated in evergreen shrubs and short-lived herbs in fully summer-dry environments, and these endemic taxa are thought to be of mainly subtropical origin and to have diversified in response to drying and other processes (e.g., increased fire frequency, mountain uplift) beginning in the Miocene. In every case except southwestern Australia, there are also mesic trees and shrubs interpreted as Tertiary paleoendemics. At more local scales, plant diversity is not particularly high in mediterranean vegetation, but there is often high turnover in composition along gradients of climate and soil. In contrast to these commonalities, there is little overlap in species or even familial composition among these five regions (Cowling et al. 1996; Dallman 1998).

The Cape Floristic Region (also sometimes considered its own Floristic Kingdom) harbors almost 6,000 endemic plant species, mainly in the fine-leaved evergreen vegetation known as fynbos. In addition to 68 percent of its species, fully 160 genera (16%) and 5 families are endemic. Diversity

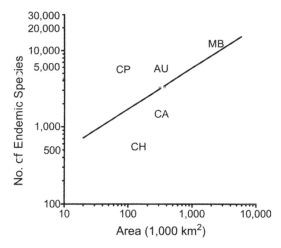

FIGURE 15. Numbers of endemic plant species in the five mediterranean climate regions as a function of their areas. (Based on data from Cowling et al. 1996)

and endemism are strongly concentrated in a few families, Aizoaceae, Restionaceae, Proteaceae, Iridaceae, Ericaceae, Rutaceae, and Orchidaceae. Most species are either geophytes or fire-killed sclerophyllous shrubs, and there are few grasses, trees, or annuals (7% annuals vs. 30% in California and 16% in Chile). Seed dispersal by ants is unusually prevalent, especially in the most species-rich groups. Some but not all of the species-rich groups are confined to nutrient-poor sandstone in the rugged and low (<2,000 m) mountains. The biogeographic origins of the species-rich groups appear to lie in other parts of Africa with varied climates but always on nutrient-poor soils (Goldblatt and Manning 2002). The eastern Cape is not seasonal enough to be truly mediterranean, and it is much lower in fire frequencies, plant endemism, and richness of evergreen shrubs and geophytes; there are more trees, a few of which are considered mesic paleoendemics. Endemic families, as opposed to species, are concentrated in the eastern Cape; they are mostly summer-flowering shrubs that are regarded as relicts of a wetter climate. Other mesic woody taxa are known in the Cape only as Miocene fossils but still persist on Madagascar (Goldblatt and Manning 2002). The Cape Floristic Region is by far the richest of the five regions in endemism corrected for area, while Chile and California fall slightly below the regression line (Figure 15).

Southwestern Australia supports around 6,000 endemic plant species, most of them found in the evergreen vegetation known as kwongan. There are many similarities to the Cape flora, including the vegetation structure, the prevalence of fire-killed shrubs, the mixture of clay and nutrient-poor sandstone soils, and the frequency of fire. The present climate and inferred climate history are also similar (Goldblatt and Manning 2002). Internal climatic gradients are somewhat milder due to gentler topography. Tertiary paleoendemics are essentially absent and may have been extirpated by aridification or fire (Cowling et al. 1996).

The comparatively vast Mediterranean Basin has over 12,000 endemic plant species. Timing and duration of the summer drought vary greatly, but summers are not as long or completely dry as in California. The region extends farther poleward than the other mediterranean climates, to 45° North, because the Alps and Pyrenees shield the basin from northerly weather. Vegetation includes evergreen shrublands (maquis and garrigue), evergreen woodland, and forest. Some of the more northerly elements of the ancestral subtropical semiarid ("Madrean") vegetation, such as oaks and pines, are shared with North America, perhaps because northerly parts of the two continents were still relatively close in the Eocene. However, the more subtropical elements are unrelated on the two continents (e.g., manzanita and ceanothus in western North America; myrtle, olive, and oleander in Europe), perhaps because the more southerly regions of the continents were farther apart at that time (Dallman 1998). Medail and Quezel (1997, 1999) identified 10 hotspots of plant endemism within the Mediterranean Basin containing 44 percent of its endemics within 22 percent of its area. In the western basin, including the Iberian Peninsula and Morocco, paleoendemism is high and diversity is related to the age of the substrate. In the east (Turkey and Greece), neoendemism is high owing to vicariance, serpentine soils, and the lesser influence of glaciation (Medail and Quezel 1997, 1999).

Central Chile harbors only around 550 endemic species. The most endemic-rich vegetation type is a sclerophyllous shrubland, or matorral, that grows less densely than Californian chaparral and does not burn frequently. Matorral grades into sclerophyllous woodland and conifer forests in wetter regions. Unlike the Cape and southwestern Australia, extensive temperate forest lies to the immediate south and contributes south-temperate genera such as *Nothofagus* and *Podocarpus*. The woody species are considered to have recruited from ancestral rainforests, reflected in the abundance of vines and bromeliads. However, among annual herb genera there is high overlap with California, amounting to 66 percent of Chilean and 41 percent of Californian genera. The herbs are considered to have evolved during the Pleistocene, although most trees and shrubs are thought to be older.

Synthesizing these patterns, Cowling et al. (1996) concluded that mediterranean climate plant diversity cannot be explained by evolutionary responses to environmental heterogeneity, because richness and endemism per unit area are highest in the Cape and southwestern Australia, where topographic, geologic, and climatic variability are modest. Instead, the ultimate driver of high mediterranean plant endemism is high speciation and/or low extinction, facilitated by stable climates and particular climatic attributes that selected for short-lived, fast-speciating lineages. Differences among the five regions, these authors concluded, relate to subtle differences in the forces acting on diversification rates. For example, in California and the Mediterranean, a considerable number of mesic (north-temperate) lineages survived and were supplemented by a modestly strong radiation of new drought-adapted lineages, but many extinctions occurred in the Pleistocene. In the Southern Hemisphere mediterranean regions, Pleistocene climates were milder because of stronger maritime influences. In Chile, this led to high survival of Tertiary lineages and less opportunity for speciation. But in the Cape and southwestern Australia, fires became frequent during and after the Pleistocene, because nutrient-poor soils selected for shrubs with fire-promoting traits such as fine, oil-rich foliage. Fire killed off the mesic Tertiary lineages, opening the way for diversification of shrubs whose postfire seed regeneration and low seed dispersal led to high speciation rates. Although some of this speciation has resulted in habitat specialization, adaptive differentiation is only a small part of the story, as witness the extraordinary functional similarity among species in the most diverse lineages (e.g., >600 species of Erica in the Cape flora, all of them small shrubs with evergreen needlelike leaves). Climate history and its effects on overall rates of speciation and extinction are therefore the "horse" driving high diversity and endemism, while ecological specialization to novel habitats is the "cart" (Cowling et al. 1996).

. . .

The comparative worldwide evidence plus the concentration of endemic plants in the mediterranean climate parts of California make it difficult to argue that either the mediterranean climate per se or its history has played a decisive role in Californian endemism. While geographic barriers may have determined the locations and timing of initial divergences and environmental heterogeneity may have helped dictate geographic patterns of speciation, these factors seem far less decisively linked to plant endemism.

With so many closely related species dominating Californian plant diversity, it has long been believed that rapid speciation in response to the mediterranean climate must be the main reason for high endemism. Yet increasing molecular and paleontological evidence suggests that the climate and flora began developing early in the Miocene, well before the mediterranean climate. Even more striking, new evidence suggests speciation rates are no higher inside than outside California. There is growing reason to think that long-term climatic stability and reduced extinction may have played a greater role, and accelerated speciation in the mediterranean climate a lesser role, than has been thought.

As in the other mediterranean climate regions, endemism in California is concentrated in relatively few lineages. It is possible that the climate has caused accelerated speciation in just those lineages, but this argument borders on circularity ("diversity is due to rapid diversification in groups with high diversity"), unless it is possible to identify key traits that allow a priori predictions of which lineages should speciate in response to a mediterranean climate. (In this respect, the analysis by Wells [1969] of postfire seed regeneration and chaparral shrub endemism remains ahead of its time.)

Paleoendemism appears to be a modest contributor to overall Californian endemism. Using the criterion of species with their nearest relatives far away, Stebbins and Major (1965) identified 177 paleoendemics, less than 10 percent of state endemics. Surprisingly, the analysis in this chapter finds that around half of north-temperate species found only in California could be considered neoendemics, in that 253 of them belong to the 50 north-temperate genera identified by Raven and Axelrod (1978) as having speciated extensively in California. However, the other 250 north-temperate species endemic to the state do not belong to the 50 species-rich genera, and this figure could be used as another rough estimate of the number of north-temperate paleoendemics in California.

Environmental heterogeneity in California is unquestionably linked to its high plant diversity but less clearly so to its endemism, as evidenced by the lesser numbers of neoendemics in the nonmediterranean parts of the state (deserts, alpine, mesic forests). At smaller scales, within the mediterranean zone, sharp climatic and edaphic gradients may foster ecological speciation into novel environments. Yet in the

Cape flora, much higher levels of plant endemism are associated with gentler gradients of topography and soil. The ultimate cause of high endemism may be a climate or a climatic history favorable to diversification, as opposed to the availability of heterogeneous environments.

There is less evidence in plants than animals for the classic phylogeographic patterns seen in many animals (discussed in Chapter 5), where sister lineages are separated by mountain ranges, plate boundaries, or present or former waterways. However, there are examples in plants of gradual allopatric divergence in response to finer-scale barriers. The genus *Limnanthes,* fragmented by the shifting locations of wetlands, the *Streptanthus glandulosus* complex, isolated to serpentine outcrops, and the *Layia* (*n* = 7) clade, separated by moderate habitat barriers, illustrate classic genetic and reproductive isolation by distance. There is usually some evidence for ecological divergence in these radiations, presumably involving adaptation to soils, temperature, water availability, and seasonal timing. With the possible exception of *Mimulus,* though, there seems to be nothing in Californian plants comparable to the dramatic adaptive radiations on islands. The contrast between the modest diversity of Californian tarweeds and the radical evolution of Hawaiian silverswords suggests that novel climates such as California's provide much less empty niche space for adaptive radiations than newly formed islands.

California's flora is rich in plants that are so weakly and inconsistently differentiated into "species" that they are challenging even for professionals to identify, and their relationships to one another are difficult to untangle even with molecular techniques. While these symptoms of a young and rapidly evolving flora make California a plant taxonomist's nightmare, they also make it an evolutionary biologist's dream. Increasingly mechanistic studies of these species groups are steadily enlarging our understanding of the roles of hybridization, chromosomal rearrangements, peripheral isolate formation, adaptation to novel environments, and other critical processes in plant diversification.

4

Animal Endemism in California

Animal endemism in California has received little attention, compared to plant endemism, and in general it is clearly lower. According to Stein et al. (2000), there are more endemic vertebrates in California than any other state: 62 full species (8% of all native vertebrates), as compared to 57 species (40%) in Hawaii and 36 species (6%) in Texas. Conservation International (2011) estimates that the California Floristic Province has 70 endemic vertebrates, accounting for 10 percent of its native fauna. Invertebrate endemism data are much scarcer (but see Kimsey 1996 and below). This chapter brings together information on Californian animal endemism that has never been synthesized before, finding that some groups are impressively endemic-rich, including amphibians, fish, and several insect and other invertebrates, and that more animal than plant studies have attempted a synthetic, cross-taxonomic understanding of the processes giving rise to new species on the California landscape. Evidence is examined for the roles of geographic barriers, the novel mediterranean climate, post-Tertiary climate stability, and ecological heterogeneity in generating animal endemism in California. Data on present-day diversity patterns, genetic variation, and biogeography are considered. The kinds and quantity of evidence are so variable for each animal group that a taxon-by-taxon approach is used, with synthesis at the end.

CROSS-TAXONOMIC COMPARISONS

For 31 native Californian animal taxa, Calsbeek et al. (2003) found that genetic breaks of the appropriate age (7–0.3 Ma, and especially 5–2 Ma) were consistently associated in space and time with the uplift of mountains (Sierra Nevada, Coast Ranges, Transverse Ranges) and the existence of present or historic waterways (Monterey Bay, Los Angeles Basin) to support the association between evolutionary divergence and geographic barriers. Examining several animal taxa (three amphibians, two mammals, one reptile, one insect, and one bird), as well as one plant taxon, each with a wide range across California, Lapointe and Rissler (2005) found that genetic variation tended to show breaks at the boundaries of ecoregions as defined by climatic differences, again suggesting that recent geographic barriers and sharp climatic gradients have produced concordant genetic structure within multiple taxa. In 75 phylogenetic lineages within 22 species of reptiles and amphibians, Rissler et al. (2006) found concordant lineage breaks across the Central Valley, San Francisco Bay, Sierra Nevada, Trinity Mountains, and Tehachapi Mountains.

The long-standing observation of a widespread, cross-taxonomic phylogeographic break associated with the Transverse Ranges was examined by Chatzimanolis and Caterino (2007), who compared patterns in the rove beetle *Sepedophilus castaneus* (Staphylinidae) to those previously found in 9 other animals, including the endemic mountain kingsnake, the California newt, and walking stick insects (genus *Timema*). The rove beetle is flightless, lives under bark on rotting logs, and may consume invertebrates or fungi. Its genetics showed a deep northwest-southeast divide between the Sierra Pelona and the San Gabriel Mountains dating to roughly 2.5 millon years, which corresponds well both with breaks identified in other studies and with the existence of a Pliocene marine embayment at the present location of the Santa Clara River drainage that was later blocked by mountain uplift.

Genetic evidence of southern ancestry followed by northerly range expansion has been found in many Californian animals, including the California thrasher and wrentit (Burns et al. 2007), titmouse (Cicero 1996), newt (*Taricha torosa*) (Tan and Wake 2002); mountain kingsnake (Rodriguez-Robles et al. 2001), ornate shrew (Maldonado, Vila, and Wayne 2001), dusky-footed woodrat (Matocq 2002), skinks in the genus *Eumeces* (Richmond and Reeder 2002), and western pond turtle (*Emys marmorata*) (Spinks et al. 2010).

Congruent genetic (phylogeographic) and interspecific (biogeo-graphic) breaks were identified by Dawson (2001) in an analysis of 41 coastal Californian marine invertebrates and vertebrates. Surprisingly, a major break was not found at Point Concepción, which has tradition-ally been considered the boundary between the "Californian" and "Oregonian" marine faunistic provinces; instead, diffuse breaks were identified in the Los Angeles region and around Monterey Bay.

MAMMALS

Mammalian species richness in the United States tends to be highest in the South and West (e.g., Texas, 161; Oregon, 150; New Mexico, 147; Arizona, 136 (Stein et al. 2000). Among U.S. ecoregions, the Sonoran-Chihuahuan Desert is the richest in mammalian species and endemism (Ricketts et al. 1999). In general, however, mammals in the temperate zone are not particularly endemic-rich. California has 17 endemic mammal species (9%), followed by Alaska (7 species, 7%) and Oregon (3 species, 2%) (CDFW 2003; Stein et al. 2000). Approximately 24 mammal species are endemic to the California Floristic Province (Table 8). Californian endemic mammals are all rodents, except for the Mount Lyell shrew (*Sorex lyalli*) and the Channel Islands fox (*Urocyon littoralis*).

The kangaroo rat genus *Dipodomys* contributes just over one-third of the state's endemic species, as well as 23 endemic subspecies (Goldingay et al. 1997). Kangaroo rats tend to be ecologically similar, primarily feeding on large seeds in the deserts, grasslands, and chaparral in South-ern and Central California. The distribution of some narrowly distrib-uted endemics suggests that, much like the "warm temperate desert" component of the flora, they expanded into the southern Central Valley and other nondesert parts of the state during the early Holocene warm period, 8,000 to 4,000 years ago, becoming trapped there by vegetation changes as a cooler and wetter climate returned (CDFW 2003).

The island fox (*Urocyon littoralis*) is the only endemic full species of mammal on the Channel Islands and has subspecies unique to each of the six largest islands. The Channel Islands spotted skunk (*Spilogale gracilis amphiala*) is an island subspecies weakly differentiated from the mainland form, with a shorter tail and wider eyes, found on the two largest islands, Santa Cruz and Santa Rosa. The widespread deer mouse (*Peromyscus maniculatus*) has distinct subspecies on all eight Channel Islands. Genetic evidence suggests relatively recent colonization for all three of these groups. Foxes and deer mice may have been transported

TABLE 8   MAMMAL SPECIES ENDEMIC TO THE STATE (S) AND/OR FLORISTIC
PROVINCE (P)

Species	Region
Mount Lyell shrew, *Sorex lyalli*	S, P
Ornate shrew, *Sorex ornatus*	P
Alpine chipmunk, *Neotamias alpinus*	S, P
Merriam's chipmunk, *Neotamias merriami*	P
Long-eared chipmunk, *Neotamias quadrimaculatus*	P
Chaparral chipmunk, *Neotamias obscurus*	P
Yellow-cheeked chipmunk, *Neotamias ochrogenys*	S, P
Lodgepole chipmunk, *Neotamias speciosus*	P
Sonoma chipmunk, *Neotamias sonomae*	S, P
San Joaquin antelope squirrel, *Ammospermophilus nelsoni*	S, P
Mohave ground squirrel, *Spermophilus mohavensis*	S
Western pocket gopher, *Thomomys monticola*	P
White-eared pocket mouse, *Perognathus alticola*	S
San Joaquin pocket mouse, *Perognathus inornatus*	S, P
California pocket mouse, *Chaetodipus californicus*	P
California mouse, *Peromyscus californicus*	P
Pacific kangaroo rat, *Dipodomys agilis*	P
Big-eared kangaroo rat, *Dipodomys elephantinus*	S, P
Heermann's kangaroo rat, *Dipodomys heermanni*	S, P
Giant kangaroo rat, *Dipodomys ingens*	S, P
Fresno kangaroo rat, *Dipodomys nitratoides*	S, P
Stephens's kangaroo rat, *Dipodomys stephensi*	S, P
Narrow-faced kangaroo rat, *Dipodomys venustus*	S, P
Salt-marsh harvest mouse, *Reithrodontomys raviventris*	S, P
California red tree vole, *Arborimus pomo*	S, P
Island gray fox, *Urocyon littoralis*	S, P

SOURCE: State data from CDFW 2003; province data estimated from Wilson and Ruff 1999.

to the islands by rafting, which was most likely 24,000 to 18,000 years
ago when the four then-united northern islands were close to the mainland. Alternatively, these animals may have been moved deliberately
by people, who colonized the islands 13,000 to 11,000 years ago and
almost certainly brought the western harvest mouse, which shows little
differentiation either on or among islands. The skunk is less likely
to have been transported by people; it appears to have colonized the
northern islands just before they were separated by rising sea levels
11,500 years ago (Floyd et al. 2011).

   To identify localities of active mammalian diversification in California, Davis et al. (2008) analyzed geographic concentrations of endemic

and near-endemic mammals. To indicate especially narrowly distributed or recently evolved species, some analyses weighted endemic species richness inversely by range sizes and/or estimated divergence times, and others included endemic subspecies. The principal hotspots thus identified were rugged regions, including the northern and central Coast Ranges, the central Sierra Nevada, the Tehachapi Mountains, and the Peninsular Ranges. The youngest endemics (probably post-Pleistocene, although absolute ages were not estimated) were found in the climatically stable outer Coast Ranges; the slightly older ones were in the more arid inner Coast Ranges. The San Francisco Bay Area and the Inyo Basin held high concentrations of named endemic subspecies. Steep gradients (outer to interior Coast Ranges), geographic barriers (San Francisco Bay), biotic crossroads (Transverse-Tehachapi Ranges), and recent physical change (postglacial colonization of the central Sierra) could all be implicated in promoting recent diversification in mammals.

BIRDS

In the continental United States bird species richness is highest in the Southwest (e.g., Texas, 472; New Mexico, 448; Arizona, 436; California, 418) but is also strikingly uniform across the country, with more than 300 species in most states (Stein et al. 2000). Endemism is low, with the outstanding exception of Hawaii, where 52 endemic species constitute 45 percent of the historically extant avifauna. The Sonoran-Chihuahuan Desert is again the richest ecoregion, and the southeastern United States is comparable in richness to the Southwest (Ricketts et al. 1999). Only two full species of birds are completely endemic to the state of California and two others to the California Floristic Province (Table 9).

The yellow-billed magpie (*Pica nuttalli*) is found in the Central Valley, coastal valleys, and adjacent foothills. It differs from the widespread North American black-billed magpie (*Pica hudsonia*) in its yellow bill and yellow streak around the eye. Based on mitochondrial DNA evidence, the two species probably diverged in the mid-Pleistocene and experienced some subsequent gene flow (Lee et al. 2003). Similar genetic breaks between Californian and other North American populations are found in a number of widespread birds. In the raven (*Corvus corax*), the split is estimated at slightly less than 2 million years old, suggesting a link with the Sierra Nevadan uplift and development of the western deserts; it is unaccompanied by any consistent morphological differences (Omland et al. 2000).

TABLE 9 BIRD SPECIES ENDEMIC OR NEARLY ENDEMIC TO THE STATE (S) AND/OR
FLORISTIC PROVINCE (P)

Species	Region
Yellow-billed magpie, *Pica nuttalli*	S, P
Island scrub-jay, *Aphelocoma insularis*	S, P
Nuttall's woodpecker, *Picoides nuttalli*	P
California thrasher, *Toxostoma redivivum*	P
Oak titmouse, *Baeolophus inornatus*	P plus southern tip of Baja
Wrentit, *Chamaea fasciata*	P plus coastal Oregon
Allen's hummingbird, *Selasphorus sasin*	P but winters to Mexico
Lawrence's goldfinch, *Carduelis lawrencei*	P but winters to Mexico
Tricolored blackbird, *Agelaius tricolor*	P but infrequently summers to Washington

SOURCE: Shuford and Gardali 2008.

The island scrub jay (*Aphelocoma insularis*), found on Santa Cruz Island, is the only insular endemic bird species in the continental United States (Delaney and Wayne 2005; Shuford and Gardali 2008). Mitochondrial DNA analyses suggest divergence from the western scrub-jay 33,000 to 150,000 years ago, with no subsequent gene flow. Because this split is much older than the separation of the four northern islands after the last ice age, in contrast to the Channel Islands endemic mammals, the jay is inferred to have gone extinct from the three smaller northern islands as rising sea levels caused these islands to shrink (Delaney and Wayne 2005).

The Nuttall's woodpecker (*Picoides nuttalli*) ranges from southern Oregon to northern Baja California in the oak woodlands of the Sierra Nevada and the Coast Ranges. It closely resembles the widespread ladder-backed woodpecker (*P. scalaris*), from which it may have diverged during the Pleistocene. The California thrasher (*Toxostoma redivivum*) is found in chapparal habitats in California and Baja California. Like the near-endemic wrentit (*Chamaea fasciata*), the California thrasher shows genetic indications of Southern Californian ancestry, followed by northerly range expansion; in contrast, the Californian population of the white-headed woodpecker, which lives in conifer forest and ranges through the Pacific Northwest, shows the reverse pattern (Burns et al. 2007). Genetic evidence in all three species suggests recent range expansion and north-south breaks across the Transverse Ranges, but more localized genetic break points are not concordant either with one another or with the breaks found for other animal taxa by Calsbeek et al. (2003) and Lapointe and Rissler (2005).

In addition to the full endemics, five species of birds are "nearly endemic" to California (Table 9). All of these require chaparral and/or coniferous-oak forest habitats, except for tricolored blackbirds, which nest colonially in Central Valley marshes (BirdLife International 2011). The San Francisco Bay Area and coastal Southern California are identified by the CDFW (2003) as having the greatest concentrations of highly range-restricted bird species in the state. Guadalupe Island is considered an additional important endemic bird area within the California Floristic Province (Conservation International 2011); it has several endemic subspecies and an extinct endemic full species (the Guadalupe caracara, *Caracara lutosa*).

An additional 61 subspecies are endemic to California, and these belong to numerous species; the species with the most endemic subspecies is the song sparrow, *Melospiza melodia,* with five (Shuford and Gardali 2008). "Nearly endemic" by the criterion of greater than 80 percent occurrence in the state are 47 subspecies, again well distributed among species—although 6 of them are wintering subspecies of the fox sparrow, *Passerella iliaca* (Shuford and Gardali 2008).

The famous California condor (*Gymnogyps californianus*), the largest land bird in North America, is sometimes referred to as a Californian endemic, but the current range of the approximately 400 wild condors encompasses Arizona, Utah, and Baja California (Mac et al. 1988; Conservation International 2011). It is considered to have survived in the western United States as the remnant of a species that ranged broadly in the New World during the Pleistocene and to have declined in the twentieth century mainly because of lead poisoning from eating bullet-killed carcasses. In 1987 the entire known population of 22 birds was brought into a captive breeding and release program.

Other birds with high conservation profiles are the federally threatened coastal California gnatcatcher (*Polioptila californica californica*), the impending listing of which led to major changes in California conservation law (see Chapter 5) and the taxonomic status of which remains debated (Zink et al. 2000); and the California spotted owl (*Strix occidentalis occidentalis*), which is less closely tied to old-growth forests than the northern spotted owl (*Strix occidentalis caurina*) but which is threatened by the increasing frequency of severe wildfires in the Sierra Nevada (Bond et al. 2010).

REPTILES

Reptiles are richest in the southernmost and warmest regions of the United States (Ricketts et al. 1999; Stein et al. 2000) and of California

TABLE 10   REPTILE SPECIES ENDEMIC TO THE STATE (S) AND/OR
FLORISTIC PROVINCE (P)

Species	Region
California legless lizard, *Anniella pulchra*	P
Blunt-nosed leopard lizard, *Gambelia sila*	S, P
Panamint alligator lizard, *Elgaria panamintina*	S
Mojave fringe-toed lizard, *Uma scoparia*	S
Sandstone night lizard, *Xantusia gracilis*	S
Granite night lizard *Xantusia henshawi*	P
Channel Islands night lizard, *Xantusia riversiana*	S, P
Sierra night lizard, *Xantusia sierrae*	S, P
Southern rubber boa, *Charina umbratica*	S, P
Forest sharp-tailed snake, *Contia longicaudae*	P
Giant garter snake, *Thamnophis gigas*	S, P
Aquatic garter snake, *Thamnophis atratus*	P
Sierra garter snake, *Thamnophis couchii*	S

SOURCE: Jennings and Hayes 1994, with modifications from Rodriguez-Robles et al. 2001; Stebbins 2003; Feldman and Hoyer 2010; and G. Pauly and R. Thomson pers. com.

(CDFW 2003). California's reptile richness (83) is well exceeded by that of Texas (149) and Arizona (100), and also by Alabama (85). Endemism is low in all states, the highest being California (5 species, or 6%), Texas (6 species, or 4%) and Florida (5 species, or 7%). Aquatic turtles are prominent in the Texan fauna, including its endemics. And Florida's reptile fauna is rich in relict sand dune species, reflecting its ancient dry climate. The majority of California's reptiles are lizards and snakes associated with the warm Mojave and Colorado Deserts (CDFW 2003).

Considering both the state and the Floristic Province, and including the recent recognition of two species (*Thamnophis couchii* and *Contia longicaudae*), there are eight lizards and five snakes endemic to California (Table 10). These numbers exclude two species that Conservation International (2011) considers Floristic Province endemics: the Cedros Island diamond rattlesnake (*Crotalus exsul*), now considered a subspecies of the red diamond rattlesnake (*C. ruber*), and the Cedros Island horned lizard (*Phrynosoma cerroense*), which has been shown to cluster genetically with desert populations outside of the state and the Floristic Province (Leaché et al. 2009). Reptiles endemic to California and/or the Floristic Province share little in terms of range or habitat (Jennings and Hayes 1994).

Near-endemics (not listed exhaustively here) include the Colorado Desert fringe-toed lizard (*Uma notata*), found also in northeastern Baja; and the California mountain kingsnake (*Lampropeltis zonata*), which is found throughout the California Floristic Province and has a disjunct occurrence in southern Washington. Five of the mountain kingsnake's seven traditionally recognized subspecies, based on color patterns, are restricted to California. Its distinctive white-black-red-black-white banding sets it apart from other kingsnakes, and it is considered by some to be a coral snake mimic. Genetic analyses indicate that the species probably evolved 15 to 5 million years ago during Miocene cooling and drying; the splits between its southern and northern clades, and further splits within the northern clade, occurred 5.7 to 2 million years ago when marine barriers divided the southwestern from the northern coastal areas and increasingly separated the coastal from the inland mountains. The color variations used to name subspecies do not match up well with the genetic patterns (Rodriguez-Robles et al. 2010).

FISH

Freshwater fish diversity is outstandingly high in the southeastern United States, especially the large and ancient river basins of the Ozark Plateau, the Cumberland Plateau, and the Atlantic Coastal Plain (Allister et al. 1986; Stein et al. 2000). Richness peaks in Tennessee (290 species), Alabama (287), and Georgia (268); the low endemism, around 4 percent, in these states is largely the result of artificial political boundaries (Stein et al. 2000). The much smaller and historically less stable river basins of the western United States support a generally depauperate fish fauna, but there are pockets of high endemism. California's 68 native freshwater fish include 20 endemics (29%), Nevada's 45 include 13 endemics (also 29%), and Oregon's 67 include 10 endemics (15%) (Stein et al. 2000). Endemism in California fishes increases to 60 percent and the total number of taxa to 113 when subspecies and genetically distinct runs of anadromous species are included (Moyle and Williams 1990; Moyle, Katz, and Quinones 2011); these runs were designated as distinct population segments under the U.S. Endangered Species Act. There are 22 full species endemic to either the state or the Floristic Province, another 6 nearly so but also ranging into the Upper Klamath watershed and/or the Modoc Plateau, and 6 completely endemic to the latter region (Table 11). There is probably considerable "hidden" endemic diversity in Californian fish, in the form of subspecies and varieties (especially in

TABLE 11   FRESHWATER AND COASTAL FISH SPECIES
ENDEMIC TO THE STATE AND/OR FLORISTIC PROVINCE

Species	Region
Kern brook lamprey, *Lampetra hubbsi*	S, P
Pit-Klamath brook lamprey, *Lampetra lethophaga*	K
Klamath River lamprey, *Entosphenus similis*	K
Blue chub, *Gila coerulea*	K
Arroyo chub, *Gila orcutti*	S
Thicktail chub, *Siphatales crassicauda*	S, P
Hitch, *Lavinia exilicauda*	S, P
California roach, *Lavinia symmetricus*	S, P*
Sacramento blackfish, *Orthodon microlepidotus*	S, P
Sacramento splittail, *Pogonichthys macrolepidotus*	S, P
Clear Lake splittail, *Pogonichthyis ciscoides*	S, P
Hardhead, *Mylopharodon conocephalus*	S
Sacramento pikeminnow, *Ptychoceilus grandis*	S
Sacramento sucker, *Catostomus occidentalis*	S, P*
Modoc sucker, *Catostomus microps*	S*
Klamath largescale sucker, *Catostomus snyderi*	K
Lost River sucker, *Deltistes luxatus*	K
Santa Ana sucker, *Catostomus santaanae*	S, P
Shortnose sucker, *Chasmites brevirostris*	K
Owens sucker, *Catostomus fumeiventris*	S
Owens pupfish, *Cyprinodon radiosus*	S
Salt Creek pupfish, *Cyprinodon salinus*	S
Delta smelt, *Hypomesus transpacificus*	S, P
California killifish, *Fundulus parvipinnis*	P
Sacramento perch, *Archoplites interruptus*	S, P
Tule perch, *Hysterocarpus traskii*	S, P
Riffle sculpin, *Cottus gulosus*	S, P
Pit sculpin, *Cottus pitensis*	S*
Rough sculpin, *Cottus asperimus*	P
Marbled sculpin, *Cottus klamathensis*	P*
Tidewater goby, *Eucyclogobius newberryi*	P
Longjaw mudsucker, *Gillichthys mirabilis*	P

SOURCE: Moyle 2002, with revisions from P. B. Moyle pers. com.

* Near-endemics to the state or province that range slightly into southeastern Oregon in the Upper Klamath and/or Pit River watersheds.

S = state; P = province; K = species entirely endemic to the Upper Klamath watershed and/or the adjacent Modoc Plateau.

minnows, family Cyprinidae) that show morphological and meristic differences suggesting local adaptation. New genetic evidence supports the opinions of earlier taxonomists that many of these taxa should be elevated to full species (P. B. Moyle pers. comm.).

The Klamath-Siskiyou region and adjacent Modoc Plateau, in north-western California to south-central Oregon, is considered second only to the U.S. Southeast as a center of fish richness and endemism (Allister et al. 1986; Stein et al. 2000). The Sacramento River system is also rich in both newly evolved and relict endemics. Death Valley is a hotspot for fish with extremely small ranges, but these are mostly at the subspecific level and include many pupfish (*Cyprinodon*) living in springs emerging from limestone (Allister et al. 1986; see also Figure 8b, above).

The depauperate yet endemic-rich Californian fish fauna is the result of long isolation with little chance to move around because of the nature of watersheds; freshwater fish can move to new regions only if they are carried there by tectonic plate movement or if the mountain barriers separating watersheds disappear. At the same time, many extinctions resulted from climatic drying and extreme Quaternary fluctuations. Among the rich ancestral Eocene fish fauna were many families that still survive in the large river basins of the eastern United States but have become extinct in the west. In the Oligocene, a diverse western fauna partly resembling the modern one developed in lowland rivers flowing westward to the coast over broad plains. The interior Snake Basin was cut off from the coastal Sacramento River basin by tectonic events around 17 to 10 million years ago. The Miocene fauna became impoverished by extinctions caused by mountain uplift and volcanism (Minckley et al. 1986).

Paleoendemic survivors of the ancient fauna, found in large and stable stream basins, include the green sturgeon (*Acipenser medirostris*), the hardhead (*Mylopharodon conocephalus*), and the Sacramento blackfish (*Orthodon microlepidotus*). The green sturgeon is a near-endemic, whose oceanic feeding grounds range northward to Alaska but which spawns only in the Sacramento, Klamath, and Rogue [Ore.] Rivers, with 90 percent of the fish coming from the Klamath. Neoendemics that speciated in response to isolation and climate change include many minnows (Cyprinidae) and suckers (Catostomidae). Most of the neoendemics inhabit large, stable rivers, and many show specialized adaptations to the mediterranean climate, such as spring spawning. The California roach (*Lavinia symmetricus*) is adapted to the warm, oxygen-poor, and generally predator-free waters of small, unstable drainages, where it has diverged into numerous locally adapted lineages that probably merit species status (P.B. Moyle pers. comm.). Desert neoendemics such as the pupfish (cyprinodontoids) are the by-products of the fragmentation of Pleistocene lakes (Minckley et al. 1986).

AMPHIBIANS

Amphibian diversity, similarly to that of freshwater fish, is highest in the southeastern United States. North Carolina (79 species), Georgia (77), and Virginia (72) exceed the richest western states, Texas (68) and California (50), although political boundaries once again lead to low (<5%) state-level endemism in the Southeast (Stein et al. 2000). Salamanders are particularly diverse in Appalachia, frogs and toads in the southeastern coastal plain (Ricketts et al. 1999; Stein et al. 2000). Amphibian richness and endemism are strikingly higher in California than in its neighboring states; Oregon has 32 amphibians and only 2 endemics, Washington has 27 species and 2 endemics, Arizona has 28 species and 2 endemics, and Nevada has 17 species and no endemics (AmphibiaWeb 2011).

With numerous recent taxonomic revisions, the state of California now has 35 endemic full species of amphibians, and the California Floristic Province has 37 endemic full species (Table 12), giving amphibians even higher proportional endemism than plants in California. This should be considered a work in progress, as new discoveries and revisions based on molecular systematics are steadily increasing the numbers of endemic amphibians.

The Klamath-Siskiyou region is noted as a U.S. hotspot of amphibian, and especially salamander, endemism, second only to Appalachia (Ricketts et al. 1999; Stein et al. 2000; Conservation International 2011; CDFW 2003). The northern Coast Ranges and the coastal fog belt are also rich in endemic amphibians, and some of the rarest endemics are found in the southern Sierra Nevada (CDFW 2003; see also Figure 8c, above).

Most Californian amphibian species are thought to be fairly old (D. Wake pers. comm.). The web-toed salamander (genus *Hydromantes,* with two endemic species) is also found in Italy and France (Conservation International 2011); in an interesting parallel to the plants that Raven and Axelrod (1978) termed "Madrean-Tethyan," its distribution suggests very ancient linkages between the Northern Hemisphere's two mediterranean climate regions. The black salamander (*Aneides*

Species	Region
Arroyo toad, *Bufo (Anaxyrus) canorus*	P
Yosemite toad, *Bufo (Anaxyrus) canorus*	S, P
Black toad, *Bufo (Anaxyrus) exsul*	S
Western spadefoot toad, *Spea hammondii*	P
California treefrog, *Pseudacris cadaverina*	P
Yellow-legged frog, *Rana boylii*	P
California red-legged frog, *Rana draytonii*	S, P
Southern mountain yellow-legged frog, *Rana muscosa*	S
Sierra Nevada yellow-legged frog, *Rana sierrae*	S
California giant salamander, *Dicamptodon ensatus*	S, P
California tiger salamander, *Ambystoma californiense*	S, P
Redbelly newt, *Taricha rivularis*	S, P
California newt, *Taricha torosa*	S, P
Sierra newt, *Taricha sierrae*	S
Scott bar salamander, *Plethodon asupak*	S, P
Del Norte salamander, *Plethodon elongatus*	P
Siskiyou Mountains salamander, *Plethodon stormi*	P
Greenhorn Mountains slender salamander, *Batrachoseps altasierrae*	S, P
Desert slender salamander, *Batrachoseps aridus*	S
California slender salamander, *Batrachoseps attenuatus*	P
Fairview slender salamander, *Batrachoseps bramei*	S, P
Inyo Mountains slender salamander, *Batrachoseps campi*	S
Hell Hollow slender salamander, *Batrachoseps diabolicus*	S, P
San Gabriel slender salamander, *Batrachoseps gabrieli*	S
Gabilan Mountains slender salamander, *Batrachoseps gavilanensis*	S, P
Gregarious slender salamander, *Batrachoseps gregarius*	S, P
San Simeon slender salamander, *Batrachoseps incognitus*	S, P
Sequoia slender salamander, *Batrachoseps kawia*	S, P
Santa Lucia Mountains slender salamander, *Batrachoseps luciae*	S, P
Garden slender salamander, *Batrachoseps major*	S, P
Lesser slender salamander, *Batrachoseps minor*	S, P
Blackbelly slender salamander, *Batrachoseps nigriventris*	S, P
Channel Islands slender salamander, *Batrachoseps pacificus*	S, P
Kings River slender salamander, *Batrachoseps regius*	S, P
Relictual slender salamander, *Batrachoseps relictus*	S, P
Kern Plateau slender salamander, *Batrachoseps robustus*	S, P
Kern Canyon slender salamander, *Batrachoseps simatus*	S, P
Tehachapi slender salamander, *Batrachoseps stebbinsi*	S
Black salamander, *Aneides flavipunctatus*	P
Arboreal salamander, *Aneides lugubris*	P
Wandering salamander, *Aneides vagrans*	P
Limestone salamander, *Hydromantes brunus*	S, P
Mount Lyell salamander, *Hydromantes platycephalus*	S, P
Shasta salamander, *Hydromantes shastae*	S, P

SOURCE: Jennings and Hayes 1994, with revisions based on Stebbins 2003; Mead et al. 2005; Kuchta 2007; AmphibiaWeb 2011; Jokusch et al. 2012; and G. Pauly, R. Thomson, and D. Wake pers. com.

flavipuncatus), inhabiting trees in coastal forests, has disjunct populations in the northern Coast Ranges and Santa Cruz Mountains that appear to have been separated since the Pliocene (AmphibiaWeb 2011).

The slender salamander (genus *Batrachoseps*) consists of 22 species, 21 of which are restricted to the state and/or the Floristic Province; the remaining one species (*B. wrighti)* is endemic to central-western Oregon. This genus consists of cryptic species with very low mobility, highly similar niches and morphology, and usually nonoverlapping geographic ranges. Genetic differentiation and reproductive isolation are nonetheless strong. Speciation in this group is believed to have begun around 30 million years ago, and tectonic plate movements may be responsible for several of the oldest phylogeographic splits (Figure 16). Species boundaries are thought to be maintained by a combination of low mobility, slow evolution of adaptive traits, and competition between closely related and nearly identical taxa, with the result being a striking pattern of "nonadaptive radiation" in California (Wake 1997, 2006). Also, the small ranges and moist habitats of many species suggest they are relicts of past wetter climates. For example, *B. campi* is found in the riparian vegetation of springs and seeps surrounded by desert terrain in the Inyo Mountains, and *B. simatus* is known from north-facing localities where riparian forest abuts exposed talus slopes in the Kern River Canyon.

The California newts (*Taricha torosa* and *T. sierra*) are closely similar sibling species and were considered subspecies of *T. torosa* until recently (Kuchta 2007). They are found in the northern and central Sierra Nevada and in the Coast Ranges plus an isolated southern Sierra Nevadan population, respectively. The two species are considered to have diverged 7 to 13 million years ago (Tan and Wake 2002; Kuchta and Tan 2006). *T. torosa* shows less geographic differentiation in the north than in the south, indicating a southern origin and northward range expansion. It appears to have colonized from Southern California to Monterey about 5 to 3 million years ago, during the uplift of the southern Coast Ranges. It then colonized northward from Monterey between 1.1 and 0.3 million years ago, following the disappearance of the large seaway that drained out from the Central Valley at Monterey Bay. *T. sierrae* also shows patterns consistent with northward expansion, as well as deeper geographic structuring than *torosa*, perhaps reflecting the rugged topography of the Sierra Nevada.

Although the frog genus *Rana* occurs worldwide, all of California's species belong to a western North American group that began diverging about 8 million years ago (Macey et al. 2001). The foothill yellow-legged

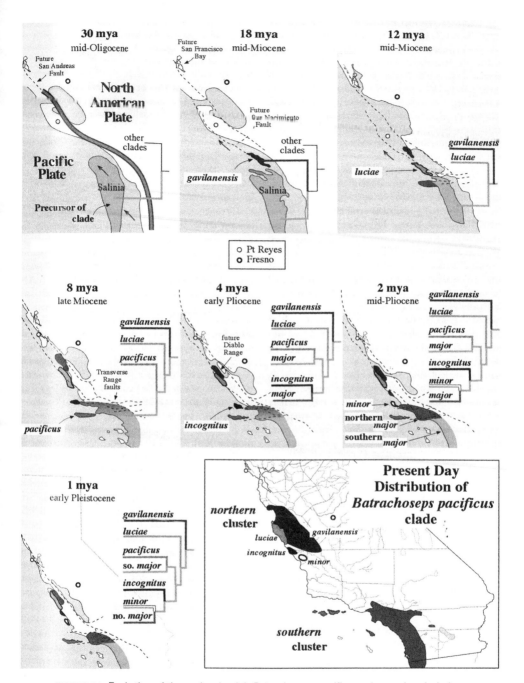

FIGURE 16. Evolution of the endemic-rich *Batrachoseps pacificus* salamander clade in relation to plate tectonic events in Southern California. Reproduced by permission from D. B. Wake, "Problems with Species: Patterns and Processes of Species Formation in Salamanders," *Annals of the Missouri Botanical Garden* 93 (2006): 8–23.

frog (R. *boylii*) inhabits nearly all mediterranean climate regions of the state, although it ranges slightly too far north into central Oregon to be a strict Floristic Province endemic. The California red-legged frog (R. *draytonii*) was once considered conspecific with the northern red-legged frog (R. *aurora*), which ranges much farther north, but molecular and morphological studies support a species break in Mendocino, where numerous other species show range limits or phylogeographic breaks (Shaffer, Fellers et al. 2004). The Sierra Nevada yellow-legged frog (R. *sierra)* in the northern and central Sierra was likewise found to be distinct from the mountain yellow-legged frog (R. *muscosa*) in the southern Sierra and southern California mountains, based on genetic splits estimated at 2.2 million years old that are concordant with behavioral and morphological differences (Macey et al. 2001; Vredenburg 2007).

The California tiger salamander (*Ambystoma californiense*) is one of the clearest instances of animal evolution in response to the mediterranean climate. It belongs to a complex of 15 recognized North American species that diverged allopatrically during and since the Pleistocene. While its close relatives breed in permanent waters, A. *californiense* has adapted to California's ephemeral waters by having a short larval period and long terrestrial dormancy. Adults travel distances up to several kilometers from vernal pools to upland hibernation sites. Matching the absence of predatory fish from its breeding pools, A. *californiense* is less aggressive than other *Ambystoma*. It is considered the most distantly related of the North American *Ambystoma* complex, likely dating to the Sierran uplift of 5 to 3 million years ago. Within the state there are geographically delimited clades that range from young and indistinct in the Central Valley to as much as 0.7 to 0.9 million years old in coastal regions (Santa Barbara and Sonoma Counties). Nonetheless, it remains reproductively compatible with other *Ambystoma*. In many parts of the state, A. *californiense* is hybridizing with Texan tiger salamanders (A. *tigrinum*) that were introduced into Californian stock ponds because their longer larval period makes them more useful as fish bait (Shaffer, Pauly et al. 2004; Fitzpatrick et al. 2010).

The western spadefoot toad (*Spea hammondi*) inhabits pools and streams in Californian foothill habitats that are even more ephemeral than the habitats of *Ambystoma californiense,* and has an extremely short larval period (<1 month). In contrast to *Ambystoma,* this short larval period is not a novel adaptation but is also found in other spadefoot toads, which are a desert group (Buchholz and Hayes 2002). The Yosemite toad (*Bufo canorus*), found in central Sierran subalpine meadows, is likewise adapted to a short growing season through traits it shares with close relatives; it is believed to be a geographically isolated derivative of *B. boreas.*

INVERTEBRATES

Most organisms on earth are invertebrates, and vastly less is known about their identities, diversity, endemism, or evolution than is the case for plants or vertebrates. Estimates of the numbers of species on earth (e.g., Mora et al. 2011) are, to a good approximation, estimates of how many million invertebrates remain to be discovered and described. Whereas most changes to California's flora and vertebrate fauna arise only from taxonomic revisions, it has been estimated that 75 percent of its insect fauna has yet to be described. Below, Californian endemism is considered first in insects as a whole, then in specific insect orders that have been well studied, and finally in other invertebrates, including arachnids (spiders and harvestmen), land snails, and freshwater crustaceans.

California has about 28,000 described native insect species and perhaps three times that number of undescribed ones (Kimsey 1996). While some sources describe California as rich in endemic insects (Conservation International 2011) and others do not (e.g., Ricketts et al. 1999), actual data are scarce. From 1940 to the mid-1980s, entomologists under the auspices of the California Insect Survey described the state's insect fauna, typically one family at a time, in monographs published in the *Bulletin of the California Insect Survey* (now available at essig. berkeley.edu). Kimsey (1996) estimates that the 36 families thus examined make up 12 percent of the state's insect fauna and that Californian endemism in these families averages 15 percent, a number that remains unchanged after some recent revisions (Table 13).

The Channel Islands contribute 80 species to the Californian endemic insect fauna (as well as 26 island endemic insect subspecies, 20 endemic spiders and harvestmen, and 12 endemic land snails). This could easily be either an underestimate of endemism due to lack of sampling on the islands or an overestimate due to inadequate sampling on the mainland (Miller 1985). In general, the island endemics are thought to be relicts of mainland taxa that became extinct during post-Pleistocene warming, just as is believed for plants. With the notable exception of the Orthoptera (below), there is little evidence for in situ speciation on the islands, not surprisingly given the short time since the most recent (middle Pleistocene) submergence of most islands as well as the high mobility of

Taxa	California Species	Endemics (%)	World Species
Grasshoppers (Orthoptera: Acridoidea)[a]			
Tetrigidae	5	2 (40)	500
Acrididae	186	94 (51)	2,000
Eumastacidae	4	2 (50)	100
Tanaoceridae	2	1 (50)	500
True bugs (Hemiptera)			
Hebridae	6	0 (0)	100
Mesoveliidae	2	0 (0)	100
Hydrometridae	1	0 (0)	80
Macroveliidae	3	1 (33.3)	100
Veliidae	10	0 (0)	200
Gerridae	7	0 (0)	200
Nepidae	3	1 (33)	200
Belostomatidae	8	2 (25)	500
Corixidae	30	4 (13.3)	1,000
Ochteridae	1	0 (0)	100
Gelastocoridae	4	0 (0)	100
Naucoriidae	10	3 (30)	800
Notonectidae	11	0 (0)	1,000
Flies (Diptera)			
Micropezidae	7	1 (14.3)	800
Anthomyidae	204	25 (12.3)	10,000
Conopidae	48	4 (8.3)	1,000
Stratomyiidae	61	3 (4.9)	2,000
Tephritida	123	19 (15.4)	5,000
Bibionidae	23	6 (26.1)	500
Sciomyzidae[b]	49	1 (2)	500
Beetles (Coleoptera)			
Platypodidae	1	0 (0)	500
Scolytidae	185	16 (8.6)	1,000
Rhipiphoridae	22	7 (31.8)	200
Bees, wasps, and ants (Hymenoptera)			
Megachilidae	200	53 (26.5)	5,000
Melectidae	15	1 (6.7)	100
Mutiliidae	28	3 (10.7)	2,000
Siricidae	14	1 (7.1)	100
Anthophoridae	9	1 (11.1)	5,000
Cephidae	5	0 (0)	100
Sphecidae	112	13 (11.6)	30,000
Pompilidae	75	3 (4)	1,000
Scoliidae	8	0 (0)	500

Chrysididae	196	18 (9.2)	2,000
Formicidae[c]	255	39 (15.3)	14,000
Dragonflies and damselflies (Odonata)[d]			
Coenagrionidae	32	2 (6.3)	1,200
Lestidae	7	1 (14.3)	500
Butterflies (Lepidoptera: Papilionoidea)[e]			
	269	*8 (3)*	*18,000*
TOTAL	1,629	284 (17)	73,750
With additions	2,232	335 (15)	107,950

SOURCE: From Kimsey 1996, with additions (italicized) and revisions from Strohecker et al. 1968[a]; Fisher and Orth 1985[b]; Ward 2005[c]; Manolis 2003[d]; Emmel et al. 1998[e]; Pelham 2008; and P. Opler pers. com.

many insects. Many of the endemic taxa are shared among the four northern islands that were joined in the Pleistocene (Miller 1985).

Edaphic endemism is rare in animals, but some cases are mentioned here and discussed in more detail below. Two Californian butterflies, the Muir's hairstreak (*Mitoura muiri*, feeding on MacNabs cypress, *Hesperocyparis macnabiana*) and the dusky-winged skipper (*Erynnis brizo lacustra*, feeding on leather oak, *Quercus durata*) are confined to serpentine because of the edaphic restriction of their host plants. Two nonendemic species (Lindsey's and Columbian skippers, *Hesperia lindseyi* and *H. columbia*) are partly restricted to serpentine in the Sierra, even though their host plants (grasses) are widespread, suggesting that in forested environments harsh soils create sunny openings that harbor relict populations of insects as well as plants (Gervais and Shapiro 1999). In the San Francisco Bay Area, the Opler's longhorn moth (*Adela oplerella*) and the Bay checkerspot butterfly (*Euphydryas editha bayensis*) are found in serpentine grasslands, despite having more widely distributed host plants (annual plantain, *Plantago erecta*, and cream cups, *Platystemon californicus*, respectively) (Powell and Opler 2009). Finally, many members of an endemic-rich family of harvestman (Phalangogidae) are confined to rocky outcrops (Ubick 1989; Platnick and Ubick 2008).

Grasshoppers, Crickets, and Relatives (Orthoptera)

Grasshoppers (Acridoidea), and in particular the family Acrididae, have among the highest Californian endemism of any animal group. An early monograph finds there are 113 endemics out of 211 total taxa

(Strohecker et al. 1968), and even after removing subspecies and correct-ing for taxonomic changes, endemism stands at 99 of 197 species, or 50 percent (see Table 13). Grasshoppers tend to be unspecialized feeders, although they sometimes associate strongly with particular plants on a local basis. They require moist, sandy soils for oviposition. In contrast to annual plants, they often have outbreaks after a series of dry years, and populations may decline following unusually wet years (Strohecker et al. 1968).

Of the endemic acridid grasshoppers, 32 are in the genus *Melanoplus* and 13 in *Trimerotropis,* with 47 and 29 total Californian species, respectively. It is unclear how these figures compare with elsewhere in the western United States. The genus *Melanoplus* has undergone a spec-tacular radiation throughout western North America, possibly even more rapid than the better-known radiations of Hawaiian *Drosophila* and African rift lake cichlids; most species are flightless and montane, suggesting that low mobility and glaciation-driven habitat change, as well as sexual selection, have promoted speciation in this group (Knowles and Otte 2000). Most endemic grasshoppers are found in mediterranean climates and are considered young species (Strohecker et al. 1968). The northern Coast Ranges support 10 ecologically similar *Melanoplus* and a number of habitat specialists, including two high-elevation and one coastal dune species. The central Coast Ranges have eight *Melanoplus* that appear to be an allopatric complex, several genera or species that appear to be climatic relicts, and several southern species that may have recently colonized. The southern Sierra Nevada has five *Melanoplus* at middle elevations and nine flightless high-altitude endemics that are thought to have speciated in situ since the Pleistocene. The northern Sierra and Cascade region has five endemics, including two *Melanoplus.* Southern California has eight endemics, nearly all of which are shallowly derived from widespread genera, including four *Melanoplus.* The Colorado and Mojave Deserts have few endemic grasshoppers, and the Central Valley has none.

On the Channel Islands, 12 of 54 species of grasshoppers and related Orthoptera are endemic. Based on distributional evidence, there has been more in situ evolution on the islands in Orthoptera than in other groups. Close relatives are often found on different islands, and in the northern islands in particular the distributions of endemic sister taxa are concordant with the islands' histories of fusion and separation. Island endemics tend to be in genera that also diversified on the mainland, notably *Cnemotettix* silk-spinning crickets, *Morsea* bush grasshoppers, and *Neduba* flightless katydids. Flightlessness is seen in 75 percent of island endemics as compared to 41 percent of the island Orthopteran fauna as a whole. In the Santa Monica Mountains on the nearby mainland, similar in size, topography, and vegetation to the islands, Orthopteran endemism is also high (9 of 70 species) and is concentrated in many of the same flightless taxa (89% of endemic species and 43% of the whole fauna are flightless). In contrast to classic theory, there is no evidence that flightlessness evolved in response to life on the islands; rather, flightlessness is common because such taxa are more likely to speciate and remain as island endemics (Weissman 1985).

Perhaps the best-studied evolutionary radiation in the Californian invertebrate fauna is that of the walking stick genus *Timema* (Crespi and Sandoval 2000; Law and Crespi 2002; Sandoval and Crespi 2008). Walking sticks, or Phasmatodea, are variously considered a suborder of Orthoptera or a separate order. *Timema* presently consists of 21 species, of which 18 are endemic to mountainous regions of the California Floristic Province and 3 have small ranges in adjacent parts of Nevada and Arizona. They feed on a variety of woody plants, including *Adenostema, Ceanothus, Quercus,* and *Juniperus.* Genetic, behavioral, and morphological studies show that this genus diversified through a process unknown in any other group, where lineages have become specialized to particular host plants in order to escape predation. Both within and between species, *Timema* may be green, gray, red-brown, or beige, and these colors may be solid or striped. The host plant used by a *Timema* species determines which color and pattern are most effective at disguising it from lizards and birds. For example, the striped pattern has evolved repeatedly in *Timema* lineages as they shifted from using broad-leaved host plants such as *Ceanothus* to needle-leaved ones such as *Adenostoma.* During speciation events, the ancestral host plants may be lost, and new host plants are added only if the insect lineage is preadapted by having the right color and pattern (Figure 17; Crespi and Sandoval 2000; Sandoval and Crespi 2008).

Timema is believed to have originated about 20 million years ago in the mountains of Southern California, as the chaparral biome began to develop. Fossil evidence shows that most of its major host plants were present by this time. Most modern *Timema* species originated 5 to 3 million years ago during the uplift of the Sierra and the Coast and Transverse Ranges (Sandoval and Crespi 2008). The southern (central and inner southern Coast Ranges, southern Sierra, desert) *Timema* clade is more basal and deeply differentiated than the northern (northern Coast Ranges

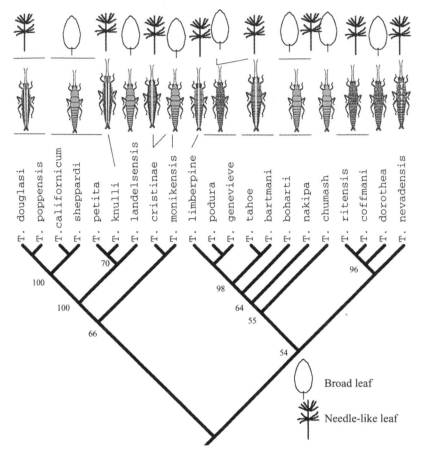

FIGURE 17. Evolution of the endemic *Timema* walking-stick insect genus with respect to host plants and cryptic color patterns. Horizontal lines separating insects indicate that those above and below have the same color patterns. Horizontal lines separating plants and insects indicate plants on which the latter feed. (Source: Reproduced by permission from Sandoval, C.P. and B.J. Crespi 2008. Adaptive evolution of cryptic coloration: the shape of host plants and dorsal stripes in *Timema* walking-sticks. Biological Journal of the Linnean Society 94:1–5.)

and Klamaths) and Transverse Range clades, indicating a northward expansion. These genetic patterns are similar to those in the California newt (*Taricha torosa*) and California mountain kingsnake (*Lampropeltis zonata*) (Sandoval and Crespi 2008). Five species of *Timema* are parthenogenetic, and these tend to be more northerly than their closest relatives, consistent with other studies linking asexuality to poleward range expansion (Law and Crespi 2002).

The genus *Stenopelmatus,* often referred to as Jerusalem crickets or potato bugs (Figure 18), has an estimated 40 to 50 species in California,

FIGURE 18. An undescribed Jerusalem cricket (*Stenopelmatus*) species endemic to Santa Catalina Island in the California Channel Islands. (Photo by David B. Weissman)

the majority of which remain undescribed. Several of these are endemic to the Southern California mountains, coast, and/or Channel Islands (Vandegast et al. 2007; Weissman et al. 2008). These insects are unusually large, reaching 7.5 cm or more, and may require more than one year to reach maturity. They are generalist feeders in a wide range of habitats. Genetic and distributional evidence suggests that the family Stenopelmatidae is a relict of past wetter climates in the Southern Hemisphere. *Stenopelmatus* appears to have undergone cryptic allopatric divergence that may be related to species-specific mating songs (Weissman et al. 2008).

True Bugs (Hemiptera)

True bugs (Hemiptera) are depauperate in California, especially the aquatic and semiaquatic families Gerridae and Corixidae, of which only 113 of North America's 415 species are found in the state. This is probably the result of the desert barrier between California and the humid tropics. However, two very narrow endemics are found only in Death Valley hot springs: *Belostoma saratogae* (Belostomatidae) and the flightless *Ambrysus funebris* (Naucoridae). Also, *Lethocerus angustipes* (Belostomatidae) is found only in Death Valley and more than 1,000 miles away in Mexico (Menke et al. 1979).

Bees, Wasps, and Ants (Hymenoptera)

Megachilid bees are solitary nectar and pollen feeders that build nests from soil or leaves, hence their common names, mason or leafcutter bees (Figure 19). A few, known as cuckoo bees, live by stealing pollen from other megachilids. Overall, the family shows high Californian endemism (94 of 186 species, or 51%). Around 124 of California's 200 species are in the tribe Megachilini, described by Hurd and Michener (1955). Megachilines are widely distributed across all ecological provinces of California, with the greatest diversity in the mediterranean climate areas. They are considered a boreal group, but several genera that are thought to have diversified in the deserts are well represented in the Californian fauna. While they are large, strong fliers, the mobility of megachilids may be limited by their fidelity to a nest site. Megachilines are unusual in being specialist pollinators, with multiple members of a bee genus or subgenus often being specialized on an entire genus of plants. For example, the largest megachiline genus in California is *Proteriades,* with 22 species, virtually all endemic to the mountains and/or deserts of Southern California and adjoining Baja California. They are specialist pollinators on the genus *Cryptantha,* having curved bristles on their mouthparts that enable them to extract pollen from the tiny flowers. There is no evidence, however, for species-level coevolution between *Proteriades* and *Cryptantha* (Hurd and Michener 1955). The megachiline genus *Atoposmia* specializes on the pollen of *Penstemon* and *Keckiella.* In both *Proteriades* and *Atoposmia,* many bee species are sympatric. Another 42 megachilids are in the tribe Anthidiini; these are also solitary, large, and showy but are unspecialized pollinators. They reach their highest diversity in the mediterranean climate areas, and 7 species are endemic to California (Grigarick and Stange 1968).

Bees in the family Andrenidae also tend to be specialist pollinators, and some are associated with vernal pool plant genera such as *Downingia* (*Panurginus atriceps*), *Limnanthes* (*Andrena limnanthis, Panurginus occidentalis*), *Lasthenia* (*Andrena submoesta, A. puthua, A. baeriae, A. duboisi,* and *A. lativentris*) and *Blennosperma* (*A. blennospermatis*). The bees forage in pool basins and nest in adjacent upland areas. Each bee visits several but not all plant species within its genus. In some areas they appear to be the principal visitors to their specialized plants, but their degree of importance to plant reproduction is unknown. As is the case for megachilines, the geographic ranges

FIGURE 19. The Californian endemic bee, *Megachile fidelis,* on a coneflower (*Rudbeckia* sp.). (Photo by Kathy Keatley Garvey)

of the bees are subsets of those of their plants, suggesting the bees are not required as pollinators, but this may reflect the highly incomplete knowledge of bee systematics and distributions (Thorp and Leong 1996).

Dragonflies and Damselflies (Odonata)

Throughout North America, most odonates seem to have fairly extensive ranges, perhaps because they tend to be good fliers and generalized predators. The few Californian endemics belong to the Coenagrionidae (pond damselflies), which are smaller, more fragile, and probabaly more sedentary than other groups. The San Francisco forktail (*Ischnura gemina*) is one of the rarest odonates in the United States, found only in the San Francisco Bay Area in small populations. It belongs to a large genus and is probably derived from the very similar and widespread *Ischnura denticollis.* The exclamation damsel (*Zoniagrion exclamationis*) belongs to its own genus. It is found in mediterranean regions of the state but flies earlier in the spring in cooler weather than most sympatric damselflies, suggesting it may be a relic of past cooler climates. The black spreadwing (*Lestes stultus*) is nearly endemic to California; it has been reported from Oregon to San Diego County. It may only be a subspecies of the widespread emerald spreadwing (*Lestes dryas*), however. Several endemic subspecies or varieties belong to the Gomphidae (clubtails), which tend to be specialized in their use of certain substrates and types of streams. The California race of the serpent ringtail (*Erpetogomphus lampropeltis*), for example, breeds on Southern Californian small streams and is separated from the other race by vast reaches of the Colorado and Mojave Deserts that lack such habitat (Manolis 2003; Manolis and Biggs 2008).

Butterflies and Moths (Lepidoptera)

Butterflies show modest endemism in California, with only 8 full species endemic to the state and a total of 27 endemic to either the state or the Floristic Province (Table 14). While there are more butterflies on the endangered species list in California than any other state, they are nearly all subspecies, many of them small-bodied and relatively sedentary members of the families Lycanidae (blues, coppers, elfins, and hairstreaks) and Hesperiidae (skippers) (Mac et al. 1988). Plant diversity might reasonably be expected to drive diversity in butterflies, since many of them are relatively specialized on particular larval host plants. However, there appears to be no geographic correlation between plant and butterfly diversity in California once the direct responses of diversity in each group to climatic variation are accounted for; this is true even when considering only butterflies known to have narrow diets and only plants known to be fed on by butterflies. The diversity patterns of butterflies appear to be shaped more by climate and topography than by plant availability (Hawkins and Porter 2003).

Despite their low endemism, Californian butterflies have been frequent subjects of genetic and evolutionary studies, far more than most other invertebrates. Below are a few examples of speciation studies involving Californian butterflies.

The endemic Muir's hairstreak (*Mitoura muiri*) is found primarily in the Coast Ranges and feeds on the Sargent's and McNab's cypresses (*Cupressus sargentii, C. macnabiana*) found mostly on serpentine soils. Its closest relative (*M. nelson*) is found from British Columbia to Southern California and feeds on incense cedar (*Calocedrus decurrens*). The distributions of *M. nelsoni* and *M. muiri* overlap in a few locations. The two species are indistinguishable based on allozymes and very similar based on mitochondrial DNA (0.2–1.1% divergence). Several

TABLE 14   BUTTERFLY SPECIES ENDEMIC TO THE STATE (S)
AND/OR FLORISTIC PROVINCE (P)

Species	Region
Sierra Nevada parnassian, *Parnassius behrii*	S, P
Gray marble, *Anthocharis lanceolata*	P
Sierra green sulphur, *Colias behrii*	S, P
Harford's sulphur, *Colias harfordi*	P
California dogface, *Zerene eurydice*	P
Rockslide checkerspot, *Chlosyne whitneyi*	S
California crescent, *Phyciodes orseis*	P
Muir's hairstreak, *Callophrys muiri*	S, P
Thorne's hairstreak, *Callophrys thornei*	P
Gorgon copper, *Lycaena gorgon*	P
Hermes copper, *Lycaena hermes*	P
Sonoran blue, *Philotes sonorensis*	P
Small-dotted blue, *Philotiella speciosa*	P
San Emigdio blue, *Plebejus emigdionis*	S
Veined blue, *Plebejus neurona*	S
Gray blue, *Plebejus podarce*	P
Gold-hunter's hairstreak, *Satyrium auretorum*	P
Sooty hairstreak, *Satyrium fuliginosa*	P
Mountain mahogany hairstreak, *Satyrium tetra*	P
Unsilvered fritillary, *Speyeria adiaste*	S, P
Avalon hairstreak, *Strymon avalona*	S, P
California giant skipper, *Agathymus stephensi*	P
Lindsey's skipper, *Hesperia lindseyi*	P
Sierra skipper, *Hesperia miriamae*	P
Columbian skipper, *Hesperia columbia*	P
Rural skipper, *Ochlodes agricola*	P
Western cloudy-wing, *Thorybes diversus*	P

SOURCE: From Pelham 2008, with additional information from K. Davenport,
J. Emmel, P. Opler, G. Pratt, and A. Shapiro pers. com.

populations on MacNab's cypress stands in the Sierra foothills resemble
M. muiri phenotypically and ecologically but are genetically more
similar to *M. nelsoni*. This complex probably evolved very recently via
host race differentiation, possibly aided by several factors: the fact that
adults mate on and around host plants, the earlier flight season of
M. muiri, and an apparent inability of hybrids to choose the host plant
on which they perform best. The low levels of divergence among populations, subspecies, and species, seen also in other California butterflies,
may reflect the rapid changes affecting the landscape during the Pleistocene (Nice and Shapiro 2001; Forister 2005). Conversely, the existence

of more distinct and localized subspecies along the coast may reflect the relative stability of the coastal fog belt (Mac et al. 1988).

The green sulfur butterfly (*Colias behrii*) is endemic to Sierra Nevadan alpine meadows, where its larvae feed on huckleberry (*Vaccinium*) species, as do those of its closest relatives. Both mitochondrial DNA and nuclear DNA show very little structuring, and the genetic evidence suggests that the species diverged around one million years ago and rapidly expanded its range during the last glaciation (15,000–120,000 years ago), when alpinelike habitats formed a broad archipelago across what are now desert regions. At that time, *C. behrii* may have hybridized with *C. meadii* from the Rockies, explaining the high DNA similarity of these species (Schoville et al. 2011).

Hybridization of two widespread copper butterflies, *Lycaides melissa* and *L. idas,* has evidently produced a new and as yet unnamed Californian endemic. The new species occupies the alpine zone surrounding the Tahoe Basin, a more extreme habitat than is occupied by either of the parental species. Genetic evidence suggests it is a fully isolated species rather than the result of gene flow or a cline. It has two novel adaptations, strong fidelity to an alpine endemic host plant (*Astragalus whitneyi*) and nonadhesive eggs that fall near the next year's growth rather than sticking to dead plant material that blows away in winter. The new habitat and host plant, as well as different wing patterns and genitalia, may help enforce reproductive isolation. This case joins growing evidence that hybridization is a more important source of speciation and novel adaptations in animals than was previously thought (Gompert et al. 2006). Another possible but unproven example of hybrid origin accompanied by host shift involves the skipper *Pyrgus communis* in the high Sierra (Fordyce et al. 2008).

Allochronic isolation, in which adaptive differences in seasonal timing lead to reproductive isolation, appears to be important for *Mitoura* (Nice and Shapiro 2001) and possibly for divergent serpentine and non-serpentine races of the skipper *Hesperia colorado harpalus* in the Sierra (Shapiro and Forister 2005). Allochronic divergence has also been observed in *Euphilotes* butterflies feeding on different species of buckwheat, *Eriogonum,* which have strongly divergent phenology that the butterflies must track. Several dozen Californian species or subspecies of *Euphilotes* feed on pollen and developing seeds of one or a few species of *Eriogonum,* the most diverse plant genus in the state. Larvae have the remarkable ability to burrow in the soil and wait as long as several years for adequate rainfall. Adult flight periods are closely synchronized to the initial flowering periods of the host plants. Where multiple subspecies occur together, they use different hosts and have different adult flight periods. The genus may have originally diversified 7 to 2 million years ago as *Eriogonum* speciated. Ongoing diversification seems to involve opportunistic adaptation to new hosts having different flowering periods, followed by reproductive isolation due to allochrony (Pratt and Ballmer 1993; Pratt 1994). Somewhat similar evolutionary patterns have occurred in the buckwheat metalmark (*Apodemia mormo*), which may represent a group of cryptic endemic species (Pratt and Ballmer 1991).

Far more diverse and less well known than butterflies are the moths, with 4,400 known species in California (see essigdb.berekeley.edu/calmoth_species_list.html). The geographic distributions of most are too poorly known to allow even crude estimates of endemism. Especially little known are the small-bodied ones referred to as Microlepidoptera, of which there are an approximately 1,800 species, including 1,400 that have been formally described (Powell and Hsu 2004). Among the few known endemics are Opler's longhorn (*Adela oplerella,* Adelidae) found in San Francisco Bay Area serpentine grasslands, and *Ethmia brevistriga* (Ethmiinae), which is widespread in the Coast Ranges and the Sierra foothills and feeds on *Phacelia distans.* The entire genus *Ethmia* uses host plants in the Boraginaceae and copes with drought through wintertime flight activity (helped by dark pigmentation) and multiyear diapause (Powell and Opler 2009).

Channel Islands endemism is low in Lepidopteran full species (26 of 670 species, or 3.3%) (Powell 1985, 1994). The Avalon hairstreak (*Strymon avalona*) is found only on Catalina Island, giving it one of the smallest ranges of any butterfly. It is beginning to interbreed with its closest relative, the gray hairstreak (*S. melinus*), which occurs naturally on the other islands and was first found on Catalina in the 1970s (Gall 1985). Two island endemic microlepidopterans feed only on the ironwood, *Lyonothamnus californicus.* Interesting evolutionary changes on the islands have been noted in some species. Three tiger (Arctiid) moths have color polymorphisms on the islands that are rare or absent in mainland populations; three longhorn (Adelid) moths are consistently smaller on the islands, an underwing (*Catocala*) is larger, and several other owlet (Noctuid) moths are darker than their mainland conspecifics. A Gelechiid moth uses *Eremocarpus* instead of *Croton* as its larval host when found on the islands. The skipper (Hesperiid) *Ocholodes sylvanoides* has a longer flight period on the islands, probably because its competitor, *O. agricola*, is absent (Powell 1985, 1994).

Arachnids

Many spiders in the primitive infraorder Mygalomorphae (tarantulas, trapdoor spiders, and their relatives) are endemic to California, especially along the southern coast. Most of these are long-lived, sedentary habitat specialists whose juveniles do not disperse aerially by ballooning on silk strands as do more advanced spiders. The trapdoor spiders *Apomastus schlingeri* and *A. kristenae* survive in mountains around the Los Angeles Basin and in a few remnant lowland sage scrub populations; mitochondrial DNA evidence shows very high genetic differentiation among these populations (Bond et al. 2006). The trapdoor spider *Aptostichus simus* inhabits dunes from northern Baja to Point Concepción, with disjunct populations in Monterey. Although not morphologically or ecologically differentiated, *A. simus* populations are highly distinct in terms of mitochondrial DNA, indicating divergences 6.3 to 1.5 million years ago concordant with changes in coastal topography (Bond et al. 2001). The Californian turret spider (*Atypoides riversi*) shows deep divergence between Coast Range and Sierran populations (Ramirez and Chi 2004).

All eight species of *Lutica*, in the Araneae (non-Mygalomorph) spider family Zodariieae, are found only in the sand dunes of coastal and insular Southern California (Ramirez 1995). Three species are endemic to a single Channel Island (*L. clementea* on San Clemente, *L. maculata* on Santa Rosa, and *L. nicolasia* on San Nicolas). These species build sand-covered web tubes that are used to detect small prey that walk across them. Juveniles do not balloon, and terrestrial migration by adults is uncommon. Like the trapdoor spiders, there are few morphological or ecological distinctions among species. Another genus of Araneae, *Titiotus* (Tengellidae), consists of 16 species endemic to California; they are found in the Sierra Nevada, the northern and southern Coast Ranges, and the Transverse Ranges in a variety of habitats, including rocky slopes and rocky grasslands and caves (Platnick and Ubick 2008).

California is the center of diversity for the Phalangogidae, a family of harvestmen (Opiliones) within the Arachnidae (Figure 20). Of the 108 North American Phalangogids, 66 (60%) are found only in the California Floristic Province, as are several entire genera. Vicariant evolution is strongly suggested by their almost completely allopatric

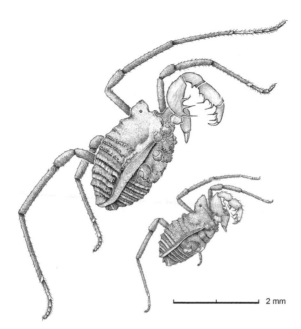

2 mm

FIGURE 20. Endemic harvestmen of the *Sitalcina* species complex, family Phalangogidae. Males of the two most common species, *Megacina cockerelli* (left) and *Sitalcina californica* (right). (Drawings by J. Speckels. Source: D. Ubick)

distributions. One group of related genera consists of large forest-dwelling species with eyes (*Sitalcina, Megacina, Enigmina*), and another consists of small, eyeless grassland species (*Microcina, Microcinella, Tularina*). These species and even these genera almost never co-occur, except in the central coast where a few species of *Sitalcina, Megacina,* and *Microcina* overlap (Ubick and Briggs 2008). Another genus, *Calicina,* more primitive than the group including *Sitalcina,* is entirely endemic to California. Members of *Calicina* are mostly small and blind as well, and live in the moist spaces under rocks in dry environments such as serpentine and sandstone in the Coast Ranges and granite and basalt in the Sierra. In only four instances do *Calicina* species co-occur, and 11 species are known from single localities. All four major species clusters within *Calicina* are found in the Sierra, suggesting this as the area of origin. Distributional patterns suggest that ancestral *Calicina* may have arrived in the Jurassic on the small oceanic tectonic plates that formed the ancestral Sierra, with subsequent speciation caused by a combination of dispersal and isolation on coastal terranes that arrived from the Eocene to the late Miocene. Small size and short life cycles may have evolved repeatedly as adaptations to open and dry habitats in an originally forest-dwelling group (Ubick 1989).

Land Snails

Snails show an impressive concentration of species richness and endemism in Appalachia that probably represents survival since the Tertiary accompanied by considerable diversification (Ricketts et al. 1999). However, terrestrial snails are strikingly rich in California, with about 200 species that are found mainly in wet and forested habitats. Almost all 52 species of the shoulderband snail genus (*Helminthoglypta*) are endemic to the California Floristic Province; although a few are endemic to the Californian deserts, their richness is highest in wet coastal and montane habitats, where they may have survived and diverged as the climate dried (Mac et al. 1988).

Freshwater Crustaceans

California's vernal pools support many planktonic crustaceans in the class Branchiopoda, including fairy shrimp (Anostraca: *Brachinecta, Linderella, Streptocephalus*), tadpole shrimp (Notostraca: *Lepidurus*) and one clam shrimp (Conchostraca: *Cyzicus*). The state currently recognizes six crustacean species as strictly confined to Californian vernal pools: *Brachinecta conservatio, B. sandiegonensis, Linderiella santarosae, Cyzicus californicus,* and *Lepidurus packardi* (Keeler-Wolf et al. 1998). Dozens of additional species remain to be described, with endemism estimated at 50 percent or higher (King et al. 1996; Helm 1998; Simovich 1998). Both species and genetic variation are highly geographically structured, showing the major role of allopatric speciation in the evolution of this group. However, ecological divergence is also important; species show adaptive differences in life cycle timing, age of maturation, and longevity that correspond to the variable physical properties of the pools (King et al. 1996; Helm 1998; Simovich 1998).

The Shasta crayfish (*Pacifasticus fortis*) inhabits cool, clear, spring-fed streams on volcanic substrates in the Pit River drainage. It exists in a handful (<10) of small and genetically distinct populations and is federally listed as endangered because of competition with two introduced crayfish species (Light et al. 1995). Another native endemic crayfish, *P. nigrescens,* is known only from nineteenth-century collections from streams in the San Francisco Bay Area. The absence of native crayfish from the Central Valley and the Sierra Nevada is puzzling considering the abundance of aquatic habitats that several alien species now occupy. The freshwater shrimp *Syncaris pacifica* occurs in small streams at low elevations in the northern Coast Ranges. It is federally listed as endangered due to the impacts of stream degradation and introduced species. Its only congener is a now-extinct Southern Californian species, *Syncaris pasadenae* (Martin and Wicksten 2004).

. . .

One of the clearest generalities that emerges from comparing animal groups is the connection between low mobility and endemism. Nearly legless slender salamanders, flightless grasshoppers, sedentary burrow-dwelling spiders, and tiny harvestmen inhabiting the undersides of rocks are among the groups that show the highest endemism in California. Moreover, these appear to have undergone primarily nonadaptive radiations, leading to complexes of ecologically similar allopatric species. In contrast, even highly mobile groups such as birds have undergone spectacular adaptive radiations not only on true islands such as Hawaii but also on ecological islands such as the Andean and Himalayan foothills, where novel montane habitat arose within tropical source areas (Price 2008). California, therefore, does not appear to be much of an "ecological island" from the perspective of animal speciation.

Geographic barriers in combination with climate change have played prominent roles in generating genetic breaks within and among animal species. Patterns of diversification in many taxa correspond in space and time with Pliocene geologic events such as the uplift of the Sierra, the Coast Ranges, and the Transverse Ranges and the opening and closing of waterways at Monterey Bay and in the Central Valley (Calsbeek et al. 2003; Lapointe and Rissler 2005; Chatzimanolis and Caterino 2007). Pleistocene and more recent climatic shifts have influenced modern endemism through the fusion and separation of the Channel Islands, the continued fragmentation of aquatic and mesic habitats, and rapid fluctuations in vegetation that may have inhibited animal specialization. While these vicariant events are critical to understanding biological diversity as it exists in California, it is less clear that they make California biologically unique. Grasshoppers, for example, may have speciated at least as much in the southern Rockies (Knowles and Otte 2000) as in California. California does not stand out as particularly noteworthy in an analysis of North American "suture zones" (i.e., hybrid zones, sister species contact zones, and phylogeographic breaks) (Swenson and Howard 2005).

Relictual endemism is apparent in Californian fish, amphibians, and spiders, many of which trace their diversification in California to the drying that began in the Oligocene. For fish and amphibians, concentrations of endemism are found in the Appalachian and Klamath-Siskiyou regions, and many other species are confined to isolated aquatic or mesic habitats—a pattern strikingly similar to that of the north-temperate (Arcto-Tertiary) flora. This simple generalization masks great complexity; some amphibians colonized California from the south, for example, and the Northern Californian hotspot for fish extends to the upper Klamath Basin, outside of both the state and the Floristic Province. The Central Valley, while poor in plants and invertebrates, is rich in endemic fish. Adaptation to the mediterranean environment, with its unstable streams, is perhaps more evident in fish than amphibians. By comparison with other regions and other animal groups, though, it seems clear that the persistence of mild, moist, and unglaciated conditions in parts of California since the Tertiary has contributed to its present-day fish and amphibian (as well as plant) endemism. The same can be said for endemism in large spiders and other narrowly distributed invertebrates that seem to be strongly associated with relatively stable coastal climates.

Considerable genetic evidence has been found for animal groups invading California from the south, for example, the California thrasher and wrentit (Burns et al. 2007), the titmouse (Cicero 1996), the newt (Kuchta and Tan 2006), the mountain kingsnake (Rodriguez-Robles et al. 2001), the ornate shrew (Maldonado, Vila, and Wayne 2001), the dusky-footed woodrat (Matocq 2002), and *Eumeces* skinks (Richmond and Reeder 2002). Equally striking is that in contrast to the presumed radiations of the southerly (Madro-Tertiary) plant taxa in California, very little speciation occurred in these groups. In one spectacular exception, the *Timema* walking sticks spread northward across California and speciated in response to the opportunity to move onto new host plants where their color patterns protected them from predation.

Animals present a number of examples of both adaptation (e.g., tiger salamanders, California roach) and preadaptation (e.g., spadefoot toads, *Ethmia* moths) to the mediterranean climate summer drought. In contrast to plants, it is difficult to identify many cases in which traits such as these appear to be linked to rapid diversification. There is also a surprising dearth of cases in which plant-feeding insects (or other animals) diversified in response to the diversification of plants, although *Euphilotes, Apodemia,* and other butterflies illustrate this phenomenon at lesser (subspecific) levels.

How can the high endemism in plants, fish, amphibians, and a few invertebrate groups be reconciled with low endemism in the majority of animals? One possible explanation linking much of the evidence in the foregoing chapters is as follows. Climatic stability is globally linked to high endemism but more strongly in species whose mobility is limited by either low dispersal capacities or narrow habitat requirements (Araujo et al. 2008; Sandel et al. 2011). During Pliocene-Pleistocene climatic fluctuations, today's mediterranean climate regions may therefore have been more significant refuges from extinction for poor dispersers (most plants, harvestmen, primitive spiders) and specialists on mild and wet habitats (fish, amphibians) than for the majority of animal groups. It is also possible that the poor dispersal capacities of plants predisposed them to the "microallopatric" mode of speciation envisioned by Stebbins and Major (1965), in which moderate Pliocene-Pleistocene climate fluctuations caused localized shifts of populations across gradients of soils and topography, leading to genetic fragmentation and speciation; more effective habitat tracking by mobile animals would lead to lower rates of speciation (cf. Wake et al. 2009). To the extent that climatic fluctuations have been more significant as causes of extinction and/or microallopatric speciation in the temperate zone while other processes have shaped patterns of diversity in the tropics and on islands, these mechanisms could help explain the unique status of mediterranean climates as hotspots of endemism for plants but not for the majority of animals—with the exception of the most water-demanding and the least efficiently dispersing ones.

—

5

Conservation Challenges in California's Endemic-Rich Landscape

California is as well known for rapid human population growth, urban sprawl, and environmental controversies as for biological diversity. Reflecting the pressures on its natural habitats, California has more federally listed species than any other state except for Hawaii, as well as more naturalized non-native plants (currently more than 1,100) (Baldwin et al. 2012). Some of the most endemic-rich natural habitats are under severe threat from habitat conversion, especially urbanizing coastal habitats from San Francisco to San Diego. Because of these pressures and because of its generally environmentally aware public and research infrastructure, California has also been an important center for developing innovative approaches to biological conservation.

This chapter reviews policy tools relevant to the conservation of the endemic-rich flora and fauna of California, including regional conservation planning, conservation easements, conservation banking, advance mitigation, and climate change adaptation. It also examines the scientific tools that feed into these policies, including ecological informatics, systematic conservation planning, and climate change science. Although none of these approaches is unique to California or explicitly targeted at endemic species, because so many of California's challenges in biological conservation arise from its mediterranean climate and the rich diversity it engenders (Table 15), this chapter takes the view that California's innovations in conservation policy and science are to some extent the by-products of its high levels of endemism. Certainly, these

TABLE 15 NUMBERS OF CALIFORNIAN THREATENED OR
ENDANGERED SPECIES AND SUBSPECIES LISTED UNDER
THE STATE AND/OR FEDERAL ENDANGERED SPECIES ACT

Group Total	Listed Taxa	Endemic Listed Taxa
Mammals	31	25
Fish	21	11
Birds	30	8
Reptiles	10	5
Amphibians	13	11
Invertebrates	34	32
Plants	281	111

NOTE: Not included are plants listed as rare, animals designated as Species of Special Concern, or the 10 federally listed runs of salmon and steelhead.

innovations will have an important effect on the future of the state's endemics.

The link between endemism and conservation was well expressed by Takhtajan (1986: xix): "The choice of territory to be protected can be made by means of very diverse criteria, ranging from the extremely practical to the purely esthetic. But no matter what the path leading to the choice of a protected territory, it is necessary everywhere, in every part of the world and with every environment, to protect from the outset the flora of those territories which represent the richest storehouse of the world's unique genetic material—namely, the endemic forms of life."

In California, as in other endemic-rich regions of the world, it is especially hard to encompass the full range of biological diversity within a few, large, publicly owned lands such as national parks. Directly or indirectly, the state's endemic richness has led to alternative ways to approach biological conservation, both scientifically (e.g., better databases and analytic tools) and in terms of policy (e.g., seeking ways to meet conservation goals on private lands).

ENDANGERED SPECIES LAWS

The cornerstone of biological conservation in the United States is the federal Endangered Species Act of 1973, which prohibits federal actions that cause increased extinction risk to species that have been listed as threatened or endangered. For listed animals but not plants, it also prohibits actions on public or private land that cause "take" (harassment, harm, death). Once a species is listed, a recovery plan is developed with the ultimate goal of bringing its population back to the point where it can be delisted.

Like many states, California also has its own Endangered Species Act (known as CESA), passed in 1984. The goals and mechanisms of CESA generally parallel those of ESA, with the exception that plants as well as animals are protected from take. In addition, California designates animals as Species of Special Concern and plants as "special status," that is, on the Rare Plants list; these species are generally included in environmental impact assessments under the California Environmental Quality Act (CEQA), which has resulted in many of the occurrence records in the California Natural Diversity Database (CNDDB) (see below).

From the ESA's earliest days, some (though by no means all) of the most controversial and influential cases concerned Californian endemics. In the early 1980s, a development proposal on San Bruno Mountain in South San Francisco was temporarily blocked because of two listed butterflies, the Mission blue (*Aricia icarioides missionensis*) and the San Bruno elfin (*Callophrys mossii bayensis*). The outcome of the controversy was a major amendment in 1982, allowing "incidental take permits" for developers in return for an approved Habitat Conservation Plan (HCP) to mitigate the impacts of an action on a listed species. Hundreds of HCPs are now approved, more in California than any other state. Beginning also in California, HCPs have evolved from being carried out by individual developers and for single species and actions to being carried out for multiple species and often on a countywide basis. In county HCPs, governmental and private participants design a comprehensive mitigation strategy for all expected impacts of future development on listed and potentially listed species. Once such a plan is approved by the U.S. Fish and Wildlife Service (FWS), which is responsible for implementing the ESA on land, the county can issue all the permits needed by developers. The intention is a better-planned and less piecemeal conservation strategy and a more streamlined process for developers (Noss et al. 1997). County HCPs in California have now merged to some extent with Natural Community Conservation Planning (see below).

Another ESA innovation that largely began in species-rich California is recovery planning for multiple species in a single geographic area that face the same threat (in all cases, habitat loss). The 1998 *Recovery Plan for Upland Species of the San Joaquin Valley* covered six listed plants and five listed animals, most of them endemic species or subspecies. The

1998 Recovery Plan for Serpentine Soil Species of the San Francisco Bay Area addressed thirteen listed plants and one listed animal, again nearly all endemics. The main recommendation of both plans was the creation of a network of protected natural areas, supplemented by restoration, research, and monitoring.

Some far-reaching ESA decisions in California have sprung from lawsuits brought by citizen groups. Among these is the 1993 listing of the Delta smelt (*Hypomesus transpacificus*) and several other endemic fish species in the Sacramento–San Joaquin Delta, which forced historic changes in water management (see below). Lawsuits to list the pika under CESA and ESA as endangered by global warming, although thus far unsuccessful, have been a national test case for using endangered species laws to regulate carbon emissions. In a broad-based lawsuit aimed at reducing pesticide use, the U.S. Environmental Protection Agency was recently sued for permitting the use of more than 400 pesticides without considering their impacts on hundreds of endangered species, including the condor, the mountain yellow-legged frog, the California tiger salamander, the green sturgeon, the San Joaquin kit fox, and the giant kangaroo rat.

NATURAL COMMUNITY CONSERVATION PLANNING

In 1991 a crisis was propelled by the proposed federal listing of the California gnatcatcher (*Polioptila californica californica*) as an endangered species when plans existed for large-scale development in the coastal sage scrub. The state legislature responded by passing the Natural Community Conservation Planning (NCCP) Act, an addition to CESA that was amended in 2003. Under this law, regional conservation plans can be established by consultation among state and federal agencies, landowners, independent scientists, and other interested parties. These plans are designed to protect a broad spectrum of diversity using a habitat-based approach, which is a more ambitious goal than the ESA's. Development is directed toward lands considered less essential for conservation, generating fees that, supplemented by federal and state support, allow more important lands to be purchased and set aside. The U.S. FWS responded to passage of the act by listing the gnatcatcher as threatened instead of endangered, permitting more flexibility in the ESA's no-take provision. The NCCP Act remains unique in the United States and is an early example of the evolution of conservation strategies away from prohibitive laws and toward collaborative approaches (Babbitt 2005).

Natural Community Conservation Plans (NCCPs) are currently under way in central coastal Orange County, San Diego County, eastern Contra Costa County, and the Coachella Valley and western Riverside County, and they are in development in Butte, Santa Clara, Imperial, Los Angeles, Mendocino, Yolo, Placer, Yuba, and Sutter Counties. The approved or nearly approved NCCPs in Orange, San Diego, and Santa Clara Counties address large numbers of endemic species (see list below), most of which lack listed status under ESA or CESA (www.dfg.ca.gov/habcon/nccp/). The species to be covered in such a plan are determined by negotiations between the project proponent (usually the county), the CDFW, and the FWS, with input from scientists. Once included in the plan, covered species are expected to be actively conserved in perpetuity (again, a stronger goal than the ESA's) and are protected from take within the county except by special permit. Design of the plans relies heavily on CNDDB data, vegetation mapping, wildlife distribution models, and expert opinion. Formal techniques for systematic conservation planning (see below) are beginning to be used.

Santa Clara Valley

Target species: 10 endemic plants (e.g., coyote ceanothus, Metcalf Canyon jewelflower, Mount Hamilton thistle); 7 endemic animals (e.g., Bay checkerspot butterfly, California tiger salamander, San Joaquin kit fox, least Bell's vireo); 4 nonendemic animals (e.g., western pond turtle, golden eagle).

Plan: Create a reserve system of 59,000 acres, including 13,000 acres of existing open-space lands. Restore and manage riparian woodland, streams, wetlands, and grasslands. Total cost ($938,000) to come from development fees (60%) and agencies (40%).

Western Riverside Multi-Species Habitat Conservation Plan

Target species: 63 plants, nearly all endemic (e.g., Munz's onion, prostrate spineflower); 5 endemic invertebrates (e.g., Delphi sands flower-loving fly, Riverside fairy shrimp); 79 vertebrates, including many endemics (e.g., Stephen's kangaroo rat, granite night lizard, San Bernardino mountain kingsnake) and nonendemics (e.g., bobcat, bald eagle, coyote).

Plan: Create a reserve system of 500,000 acres, including 346,000 acres of existing public lands. In addition to land acquisitions from willing sellers, conservation banks on private lands will contribute to the network. Cost of land acquisition estimated at $812,000,000, and annual management at $10 million.

San Diego County Multiple Species Conservation Program

Target species: 43 plants, nearly all endemic (e.g., San Diego ambrosia, San Diego barrel cactus, San Diego button celery, San Diego goldenstar, San Diego mesa mint, San Diego thornmint); 4 invertebrates of which 3 are endemic (Riverside fairy shrimp, San Diego fairy shrimp, Thorne's hairstreak butterfly); 34 vertebrates, including some endemics (e.g., California red-legged frog, San Diego horned lizard, Coastal California gnatcatcher) and many nonendemics (e.g., Canada goose, bald eagle, mountain lion).

Plan: A 172,000-acre reserve system, divided between three planning sectors of the county (www.sdcounty.ca.gov/dplu/mscp/). In addition to land acquisitions from willing sellers, conservation easements on private lands will contribute to the network.

The long-term effectiveness of NCCPs remains unknown. Observers point to many obstacles in the realms of shaky political support, bureaucratic delays, and inadequate funding (Alagona and Pincetl 2008). Science-related shortcomings include piecemeal surveys, inadequate knowledge of the needs of rare species, poor monitoring, and conflicts between the management needs of different species. Nonetheless, one recent assessment concludes that the NCCP approach is at least successful in terms of increasing the amount of land set aside for conservation. The San Diego County Multiple Species Conservation Program (MSCP), which is both a county HCP and an NCCP, has been carried out separately for three areas of the county with roughly equal amounts of private land and the same focal species. In the southern planning area, where the plan has been implemented for 10 years, there has been nearly five to ten times the conservation land acquisition (12,914 hectares) as in the northern (1,904 ha) and eastern (2,467 ha) planning areas where the plan has not yet begun. The loss of natural lands has been nearly identical in the three planning areas, but species of concern have received substantially more conservation in the southern area. Both public and private stakeholders appear to perceive benefits that motivate them to participate. Among the important remaining goals are better linkages between core habitats and more resources for management and monitoring (Underwood 2011).

CONSERVATION EASEMENTS

On private lands conservation is increasingly being accomplished by conservation easements, or legally binding agreements to restrict development so as to benefit the environment. Nonprofits known as land trusts buy or are given these easements, own the relinquished development rights, and must monitor easement properties for compliance with easement terms. Landowners can still use their property in specified ways, but the restrictions spelled out in the easement are permanent and accompany the land title when it changes hands. The concept is over a century old and is applied in other countries (in Europe they are called conservation covenants), but until recently easements could be held only by governments. Massachusetts (1956) and California (1959) were the first states to adopt laws enabling land trusts. Changes to the tax code and other laws in 1980–81 brought the modern land trust movement into being, and further legal changes have enhanced the incentives for conservation easements (Airey 2010).

Conservation easements are perceived as an economical and private property–friendly way to do conservation (Merenlender et al. 2004). They are less expensive than buying land for conservation and leave the burden of land management with the property owner. Given California's development pressures, high land costs, and strong environmental movement, it is not surprising that the number of land trusts and acreage under easements are the highest in the nation (Merenlender et al. 2004). One of the oldest land trusts is the Sempervirens Fund, founded in 1900 to protect the coastal redwood forests. The Marin Agricultural Land Trust, founded in 1980, was the first to focus on agricultural lands. The Peninsula Open Space Trust in the San Francisco Bay Area is one of the state's largest, protecting over 60,000 acres in three counties. Also in California, the Nature Conservancy began in the 1980s to use easements as an alternative to owning and managing its own lands; it is now the world's largest land trust (Kiesecker et al. 2007).

Most land trusts concentrate on open space preservation, and the efficacy of conservation easements for conserving biological values is little known (Merenlender et al. 2004; Rissman and Merenlender 2008). Conservation easements held by the Nature Conservancy are an exception, because they are usually part of a biologically informed regional conservation plan. Of 119 easements acquired by the Nature Conservancy from 1984 to 2004 in eight states, Kiesecker et al. (2007) found that 96 percent had well-identified biological targets, 84 percent were within ecological priority areas identified by regional planning, and 79 percent were adjacent to other protected lands. The Nature Conservancy has been active in using easements on ranchlands to protect Californian vernal pools, the habitat of many rare and endemic species (Rissman et al. 2007). However, biological surveys and ecological monitoring remain rare even on Nature Conservancy easements (Rissman et al. 2007).

CONSERVATION BANKING

Conservation banks are private lands managed to meet mitigation requirements concerning rare species. They evolved from and are similar to mitigation banks, which are managed for wetlands. Land with ecological value is placed under conservation easements, and the easement holders are granted credits for the land's conservation value by state and federal agencies. These credits are sold to developers, currently at $3,000 to $125,000 per acre, as a way to offset impacts on species protected under state or federal law. The idea is to avoid piecemeal mitigation, in which each development project results in one small piece of land set aside, and to channel mitigation money into creating larger and more ecologically valuable reserves. At the same time, developers benefit from a quick and flexible way to meet their regulatory obligations. The first policy allowing this form of conservation credit trading was created in California in 1995, and the first conservation bank, Coles Levee Ecosystem Preserve, covered the San Joaquin kit fox, the Tipton kangaroo rat, and the blunt-nosed leopard lizard. In 2003 the U.S. FWS implemented a policy modeled after California's. California still has more conservation banks (30 of 76) and a more active conservation credit market than any other state (Fox and Nino-Murcia 2005). Credits are being sold for endemic species in many of the state's currently active conservation and mitigation banks (www.dfg. ca.gov/habcon/conplan/mitbank).

Stillwater Plains Mitigation Bank (Shasta County)
  Valley elderberry longhorn beetle.

Bryte Ranch Conservation Bank (Sacramento County)
  Vernal pool fairy shrimp, vernal pool tadpole shrimp, valley elderberry longhorn beetle.

Dolan Ranch Conservation Bank (Wildlands, Inc., Colusa County)
  Giant garter snake.

Alton North Conservation Bank (Sonoma County)
  California tiger salamander, Burke's goldfields, Sonoma sunshine.

Desmond Mitigation Bank (Sonoma County)
  Sebastopol meadowfoam.

Elsie Gridley Mitigation Bank (Solano County)
California tiger salamander.

Haera Wildlife Conservation Bank (Wildlands, Inc., Alameda and San Joaquin Counties)
San Joaquin kit fox.

Hale Mitigation Bank (Sonoma County)
California tiger salamander, Sebastopol meadowfoam, Sonoma sunshine.

Hazel Mitigation Bank (Sonoma County)
California tiger salamander.

Liberty Island Conservation Bank (Wildlands, Inc., Yolo County)
Delta smelt.

North Suisun Mitigation Bank (Wildlands, Inc., Solano County)
California tiger salamander.

Ohlone Preserve Conservation Bank (Alameda County)
Red-legged frog, Alameda whipsnake, California tiger salamander.

Swift/Turner Conservation Bank (Sonoma County)
Sebastopol meadowfoam, Sonoma sunshine, Burke's goldfields, California tiger salamander.

Agua Fria Multi-Species Conservation Bank (Merced County)
San Joaquin kit fox, California red-legged frog.

Kern Water Bank Authority Conservation Bank (governmental, Kern County)
San Joaquin kit fox, Tipton kangaroo rat, blunt-nosed leopard lizard.

Palo Prieto Conservation Bank (Kern and San Luis Obispo Counties)
San Joaquin kit fox.

Chiquita Canyon Conservation Bank (governmental, Orange County)
California gnatcatcher.

Cajon Creek Habitat Conservation Management Area (San Bernardino County)
San Bernardino kangaroo rat, Santa Ana woolly-star, slender-horned spineflower, and 21 other threatened and endangered species.

Some conservation banks are created as part of a regional plan such as an NCCP, while others are put forward either by specialty for-profit companies or individual landowners. Quality control remains an unsolved problem in the latter cases. Conservation banks must be approved by the California Department of Fish and Wildlife, the U.S.

Fish and Wildlife Service (if there are federally listed species), and the U.S. Army Corps of Engineers (if there are wetlands), but these agencies lack the resources to assess the growing number of small private conservation banks. Conservation banks may be most effective when carried out as part of an NCCP or other regional plan, but most conservation in NCCPs is currently done by outright land acquisition.

ADVANCE MITIGATION

Transportation development is one of the largest ongoing environmental impacts in California, ranging in scale from widening local roads to the proposed construction of a high-speed rail link across the Central Valley. Problems owing to piecemeal environmental planning and mitigation for these projects have been estimated to cost the state $59 million per year (Thorne, Huber et al. 2009). The results of transportation mitigation are all too often small, scattered, and poorly monitored parcels. Recently, the state has begun to support the development of new analytic and policy tools for better solutions. The analytic approach involves informatic tools that allow queries by district, county, or watershed of data on biodiversity, the potential biological impacts of the next ten years of funded transportation projects, and the cumulative impacts and mitigation needs for the next ten years. Currently, the nearly 1,000 planned projects affect 11,000 hectares and intersect the potential ranges of 132 vertebrates and 248 plants with federal or state status (Thorne, Huber et al. 2009). On the policy side, efforts are being made to allow collective mitigation to be started even before individual projects are completed. The goals are aggregated and effective reserves, streamlined environmental review, and better coordination with other regional planning. In 2010 the Department of Water Resources launched a pilot project for a similar approach to mitigating water projects. Advance mitigation is also being developed in Florida and elsewhere (Thorne, Huber et al. 2009; Huber et al. 2010).

BIODIVERSITY INFORMATION

Science-based conservation requires good biological information accessible to a wide range of users and applications, and California has developed some relatively advanced tools for handling and using its biodiversity data. Like all states, its species occurrence information is housed in a natural heritage database, part of the nationwide system designed by the Nature Conservancy in the mid-1970s (see Chapter 1). The basic units are element occurrences, or geographic locations of well-documented observations of "conservation elements," which include state and federally threatened and endangered species, other species determined by experts to deserve inclusion, and special-status habitats. Element occurrences are derived from museum records, published literature, and environmental reporting by consultants, researchers, agency biologists, and conservation groups. The California Natural Diversity Database, managed by the CDFW, currently contains around 60,000 occurrences of over 2,000 plants and 900 animals, with hundreds of occurrences added or updated each month.

The state works closely with the California Native Plant Society (CNPS) to determine the conservation status of native plants. The California Rare Plant Ranks list (formerly known as the CNPS List; www.rareplants.cnps.org/ Inventory) is continually updated by the CNPS lead rare plant botanist in collaboration with the CNDDB and a statewide network of hundreds of agency, academic, and private botanists. Plants may be ranked as probably extinct (1A), rare or endangered (1B), rare or endangered in California but possibly common elsewhere (2), needing more information (3), or needing to be watched (4). Of the 1,600 currently listed vascular plants, more than 1,000 are in rank 1B (rare or endangered), consisting of over 90 percent of endemics. The CNPS Rare Plant Program and the CDFW share data on the distribution, endangerment status, and ecology of rare plants. The 1,600 vascular plants on lists 1B and 2 must be considered in environmental reporting under the California Environmental Quality Act.

The CNDDB currently tracks around 900 animal taxa on the state Special Animal List, which includes federal- or state-listed threatened and endangered taxa, state Species of Special Concern, and species being considered for those designations. Animal species, subspecies, or distinct populations may be included as Species of Special Concern if they are believed by state biologists to be declining, extirpated, or highly vulnerable to decline or extinction but are not yet listed by the state as threatened or endangered. Given the scarcity of data, the Species of Special Concern designation may indicate that the animal is generally known to be rare or that it occupies threatened habitats such as coastal wetlands, coastal sage scrub, vernal pools, arid scrub in the San Joaquin Valley, or riparian habitat. Although not formally protected, Species of Special Concern are required to be considered under CEQA. The Species of Special Concern list currently contains around 750 animals and is under revision (mammals were last revised in 1986, reptiles and amphibians in 1994, and fish in 1995).

The CNDDB has several features that allow geographic queries: Quick Viewer returns lists of species present in a selected area; RareFind gives detailed maps and reports on CNDDB occurrences and is used heavily by agency scientists and consultants; and the Biogeographic Information and Observation System is an online mapping tool that houses a wide array of spatial environmental data sets. A mapping tool known as Areas of Conservation Emphasis displays species richness and rarity hot spots (see Chapter 1) for the entire state and for individual regions.

Vegetation maps until recently have been fairly crude and thus of limited value for conservation efforts aimed at rare and endemic species. However, a new nationwide system of classifying and mapping vegetation has been developed in California (in part) and has been adopted by the state's Vegetation Classification and Mapping Program; it is being carried out region by region when funding allows (for details of the method and an early test application in Napa County, see Thorne et al. 2004). Compared to previous mapping, it identifies a significantly greater number of vegetation types and larger and more accurate units of rare vegetation types, although the ability to predict rare species is still not very high (Thorne et al. 2011).

SYSTEMATIC CONSERVATION PLANNING

GAP Analysis

As a systematic approach to improving the network of conserved lands, a geographic tool known as GAP analysis uses map layers for land cover, landownership, plant communities, and predicted animal distributions to identify lands that support unprotected species and communities (Scott et al. 1993; Davis et al. 1991). A GAP analysis for California (Davis et al. 1998) has been used by the state, counties, and nonprofits for regional planning, and the biogeographic data it generated (www.biogeog.ucsb.edu/projects/gap) have been used to model fire regimes, sudden oak death, dispersal corridors, climate change impacts, and land acquisition strategies (Davis et al. 1994, 1995; Stoms 2000; Stoms, Borchert et al. 1998; Stoms, McDonald et al. 1988). The GAP approach is generally better suited for widespread species and communities than for localized species such as the majority of endemics, and it seldom includes special habitats such as serpentine or limestone soils.

GAP models have been used to depict the distributions of endemic shrubs and trees with fairly large ranges, however, and there have been attempts to model rare endemics using a combination of GAP data and CNDDB rarity information (e.g., Davis et al. 2006).

GAP-style analyses have also been employed at global scales to gain a general sense of worldwide conservation threats and priorities. For the mediterranean biome, Underwood et al. (2008, 2009) compared the percentage protection of potential natural vegetation types, stratified by elevation, for the 39 terrestrial ecoregions within mediterranean forest, woodland, and scrub biomes. Overall land protection was 4.3 percent, ranging from around 1 percent in Chile and the Mediterranean Basin to around 10 percent in California, the Cape, and Australia. High elevations and shrublands were better protected than low elevations, woodlands, grasslands, or forests. Conversion to urban or high-intensity agricultural land use was 30 percent overall, ranging from 17 percent in California to 37 percent in Australia, and is most severe at low elevations.

Core-Corridor Analysis

The core-corridor or ecological network approach to designing conservation reserve systems focuses on wide-ranging animals. The mountain lion (*Puma concolor*) is a natural focal species in California because of its large area requirements (e.g., Thorne et al. 2006; Huber et al. 2010), and pioneering attempts were made to design and implement corridors for the mountain lion in urban Southern California (Beier and Noss 1998; Morrison and Boyce 2009). In a computational approach called least cost corridor analysis, land cover data are used to identify large "core" areas of natural vegetation, and "corridors," or linkages, among them are identified with an algorithm that minimize distance traveled weighted by the quality of the lands crossed (where land cover types are qualitatively ranked by their suitability for animal dispersal). This approach is being used by the Essential Habitat Connectivity Project (www.dfg.ca.gov/habcon/connectivity; Spencer et al. 2010), a joint project of the CDFW and other agencies to meet legal mandates for wildlife conservation, transportation mitigation, and climate change adaptation. As a result, several hundred priority locations have been identified for land acquisition or improved road crossings. Surprisingly, little is yet known about the actual efficacy of corridors for wildlife (e.g., Hilty and Merenlender 2004), but genetic markers are beginning to help us determine whether they actually reconnect fragmented animal populations (e.g., Ernest et al. 2003).

Corridors for large and wide-ranging animals may carry opportunity costs in terms of other conservation goals (Morrison and Boyce 2009), including the conservation of small and less mobile plants and animals such as most endemic species. One analysis found that a core-corridor network designed for mountain lions in the central Coast Ranges also protected other elements of biodiversity (redwoods, serpentine soils, steelhead) but missed the majority of endemic vertebrate populations (Thorne et al. 2006).

Reserve Design Algorithms

New tools for conservation planning known as reserve design algorithms allow multiple objectives to be optimized, including representative vegetation, rare species, and connectivity (e.g., Moilanen et al. 2009). The algorithms use maps that depict a planning region as a set of polygons, each of which contains variable quantities of desired species and vegetation types. The user sets targets, such as desired percentages of each vegetation type, rare species occurrences, and minimization of edge-to-area ratios. The algorithm seeks to meet these targets while also constraining the cost of the network in terms of land or money. Since the optimization process is inexact, the model is run multiple times, resulting in different solutions; output typically includes both the solution that meets the objectives most fully and a score for each land parcel representing how often it appears in alternative solutions. Thus far, these powerful but complex and data-intensive models remain more of a research subject than a practical tool. In California, they have been used to evaluate regional conservation plans (Church et al. 1996), marine reserve design (Klein et al. 2008), and mitigation for road impacts (Huber et al. 2010). The Bay Area Open Space Council recently released a plan for completing the conservation lands network in the nine-county San Francisco Bay Area by adding lands supporting currently unprotected rare and endemic species; a reserve design algorithm was used to construct a public database that identifies essential, important, fragmented, and "other" lands for conservation. Reserve design algorithms are also being used to help identify a set of parcels meeting the conservation objectives of the state Department of Transportation for mitigation in the Elkhorn Slough area of Monterey

County and in the Pleasant Grove area of the northern Central Valley (Thorne, Huber et al. 2009).

MARINE AND AQUATIC CONSERVATION

Marine protected areas (MPAs), or zones completely set aside from fishing, are a relatively new approach to fisheries management that may outperform the traditional approach of setting maximum harvest quotas. The goal is for fished species not only to survive securely within MPAs but also to export their offspring to surrounding areas where they can be harvested. California has been a leader in both using state-of-the-art tools to design a network of MPAs (e.g., Klein et al. 2008) and implementing MPAs under the 1999 Marine Life Protection Act (MLPA) in the face of considerable political controversy (see www.dfg. ca.gov/mlpa/ for a summary of the MLPA and a status update on the current MPA network).

Freshwater conservation in California has yet to receive the same degree of systematic attention as marine conservation. Fully 83 percent of the species in California's endemic-rich freshwater fish fauna are extinct or declining, a situation that is steadily worsening (Moyle, Katz et al. 2011). Water withdrawals, dams, deteriorating water quality, and exotic species introductions are the severest threats, together with climate change (Moyle, Kiernan et al. 2011). A multitiered conservation strategy has been proposed in which the most critically endangered species would be listed under the federal ESA, clusters of imperiled species would be managed collectively, a system of aquatic diversity management areas delineated, and restoration strategies developed at watershed and ecosystem scales (Moyle and Yoshiyama 1995). Lacking such a systematic approach, the state continues to experience periodic crises over the conflicting water needs of endangered fish, agriculture, and ever-growing urban demand. The most influential such clash began in the mid-1990s when several fish species endemic to the Sacramento–San Joaquin Delta were listed as federally endangered. Their listing triggered a realignment of water policy known as CALFED, a bold experiment in restoration and collaborative management that had strong scientific support but ultimately lacked adequate political backing; since its demise, decision making on Delta water issues has reverted to the courts, while the fish continue to disappear (Service 2007). Other multiagency efforts to restore native fish are now playing out in the San Joaquin River, the Modoc Plateau, the Klamath Basin, and elsewhere.

CLIMATE CHANGE SCIENCE AND POLICY

Current projections for the California climate over the next century include higher temperatures (1.5°–4.5°C), especially in the interior; a large decline in winter snowpack; modest (10–20%) changes in total precipitation; and a possible decline in coastal fog (Hayhoe et al. 2004; Cayan et al. 2008). Natural vegetation is expected to be especially affected by more frequent extreme heat events and more frequent and severer fires (Hayhoe et al. 2004). Detailed forecasts for Californian vegetation include a retreat in alpine and subalpine forest; replacement of conifer forest by mixed evergreen forests; and expansion of grass-lands at the expense of woodlands and shrublands, largely in response to a significant (10–15%) increase in the annual area burned (Lenihan et al. 2008). Some analyses suggest that mediterranean regions will lose more species as a result of climate change than any other biome, due to high sensitivity to multiple climate-related threats and their interactions (Klausemeyer and Shaw 2009). Climate change impacts on biodiversity in California may be buffered to some extent by rugged topography and the existence of contiguous land in the cooler (northerly) direction; the regions with the least land conversion and the most land protection, namely, mountainous areas, also happen to be the most climatically stable. These mitigating factors are absent in Australia, making it poten-tially the most vulnerable part of the mediterranean biome (and the world) to loss of biodiversity from climate change (Klausemeyer and Shaw 2009).

Evidence that warming is already having ecological impacts has come from comparisons of past and present species distributions (Parmesan 2006), and a substantial number of these studies have taken place in California. One of the first was a recensus in 1993–96 of hundreds of sites, most in California, where the checkerspot butterfly (*Euphydryas editha*) had been collected in 1860–1986. Below 2,400 meters, more than 40 percent of populations were extinct even where the habitat was still suitable, while above that elevation only about 15 percent were extinct, resulting in a mean upward shift of 105 meters in the butterfly's distribution (Parmesan 1996). Resampling of a transect running from 244 meters in the Coachella Desert to 2,650 meters in the Santa Rosa Mountains showed that woody plants have shifted upward an average of 65 meters since 1977, due to upward retractions of their lower mar-gins as opposed to expansions at their upper limits (Kelly and Goulden 2008). In Yosemite National Park, over half of 28 mammal species have shifted upward by an average of 500 meters over the past 100 years, due both to expansions of lower-elevation species at their upper margins and retractions of high-elevation species at their lower margins (Moritz et al. 2008). In the Siskiyou Mountains of southern Oregon, plant community composition has shifted over the past six decades, with present-day communities having lower representation of species of northerly biogeographic origin, more species with drought-adapted traits, and a closer resemblance to communities on warm south-facing slopes as compared to the same locations in the period 1949–51 (Damschen et al. 2010; Harrison et al. 2010) All of these studies found that biological changes matched predictions based on observed climate change and were not explainable by habitat alteration or other causes.

Species distribution models are the most widely used tool for assessing the future of biodiversity under climate change. Species' current distributions are modeled as functions of present climate, and then the distributions of those same combinations of climatic conditions in the future are predicted using models of climatic change. Species are predicted to become extinct either if their present climates do not exist in the future or if their future predicted distributions do not overlap with their current ones (assuming a worst-case scenario in which dispersal is too slow to track climate change). Such models can be used to examine the adequacy of existing protected areas networks and the optimal locations for additional reserves and corridors. Leading work of this kind has come from analyses of the future of the Cape flora in South Africa (e.g., Hannah et al. 2005). In California, models predict that at least 66 percent of endemic plants will experience range reductions of 80 percent or more, and assemblages will be dismantled as species move in varying directions, mostly toward the Klamaths, the high Sierra Nevada, or the coast (Loarie et al. 2008). Similar predictions were found in models for Californian land birds (Wiens et al. 2009). Methods, assumptions, and uncertainties of such models are discussed by Wiens et al. (2009).

Downscaling of climatic and species distribution models is at the forefront of making climate change predictions that are useful to land managers and other decision makers. Localized climatic gradients caused by slope, aspect, elevation, and proximity to the coast may create significant deviations from the predictions of coarser models. For example, in flat terrain a species might have to move 6 kilometers per year to keep pace with a 3°C per century temperature rise, whereas in rugged terrain the same species might have to move only 60 meters per year (Loarie et al. 2008). Preliminary downscaled models for the San Francisco Bay Area in the next century suggest that of 500 protected areas, only 8 of the largest ones will still experience temperatures within their current ranges; however, changes will be diminished close to the coast because of the buffering effect of the steep coast-to-interior temperature gradient (Ackerly et al. 2010).

California has unquestionably been a leader in anticipating climate change and creating science-based policies to anticipate and respond to it (Hayhoe et al. 2004; Cayan et al. 2008; Franco et al. 2008). The centerpiece of these efforts is the state's 2009 California Climate Change Adaptation Strategy (www.climatechange.ca.gov/adaptation). Led by the California Natural Resources Agency, this multiagency policy summarizes the best-known science on climate change impacts and recommends how to manage resources in light of them. In the area of biodiversity conservation, the core goal articulated by the strategy is a well-connected system of large reserves that will enable wildlife to shift their distributions in response to expected changes in vegetation. The California Department of Fish and Wildlife and its partners have been directed to make this reserve system a central part of their conservation planning.

Perhaps of greater direct relevance to endemic species, efforts are also under way to assess the vulnerability of rare species to climate change, with possible application to prioritizing species for listed status or lands for acquisition. Scientists working with the CDFW are conducting exploratory assessments of hundreds of rare species, using a combination of species distribution modeling and the Climate Change Vulnerability Index developed by NatureServe (www.natureserve.org/climatechange), which considers attributes such as life history and niche traits, sizes and locations of geographic ranges, and the existence of barriers to dispersal. Preliminary results suggest a high degree of climate change vulnerability for the majority of rare plant species examined.

. . .

Conservation in California and (to a large extent) the other mediterranean climate regions confronts the need to protect large numbers of rare species—widely scattered across the landscape and especially abundant in expensive coastal real estate—from the pressures of high human population densities and rapid development. No single conserved area, however large, could encompass more than a small fraction of California's native species or vegetation types. Climate change will worsen this already difficult situation through increasingly severe wildfires and droughts and by intensifying pressure for large-scale water storage and alternative energy development.

California has responded to these challenges in diverse ways. The wide use of conservation easements and conservation banking in California has arisen, in part, from the realization that the state's species and landscapes are too diverse to be conserved only on public lands. Development of improved tools for accessing and analyzing biodiversity information has also been significant in California. Legal mechanisms have been created to channel private developer dollars into establishing science-based regional networks of conservation lands. Advance mitigation by the Department of Transportation and the Department of Water Resources and their partners are attempts to replace piecemeal approaches with science-based planning. The biodiversity elements of the state's Climate Change Adaptation Strategy also represent a commitment to forward-thinking, landscape-scale planning, although with connectivity for large animals being a more prominent goal than rare species conservation.

Among the current proposals for coping with climate change and oil scarcity in California are a massive solar energy complex in the Mojave Desert, wind energy installations in a number of botanically rich localities, and the transvalley high-speed rail link, which may create an impenetrable barrier blocking the west-east dispersal of plants and animals. Even in the realm of "conservation values," the preservation of rare plants and animals seldom prevails over goals that have more direct benefits to humans.

6

Synthesis and Conclusions

Plant endemism does not appear to stimulate animal endemism. One of the striking results from considering Californian plant and animal endemism is the absence of an apparent link between the two. The few animal groups that show very high endemism in California, comparable to the level seen in plants (i.e., slender salamanders, acridid grasshoppers, megachilid bees, phalangogid harvestmen), are not highly specialized on plants as resources. In contrast, relatively low endemism is seen in animal groups known to be highly specific to particular host plants (e.g., butterflies and most other herbivorous insects). Walking sticks in the genus *Timema* are a fascinating and unusual case of rapid evolution in an insect group stimulated by interactions with host plants and predators; however, its host plants are a handful of common shrubs and trees, and the predators are generalist birds and lizards, so *Timema* diversification is not a response to diversification in either plants or other animals. The megachilid bees deserve more study, since their specialism on particular plants as pollen sources could have some link to their high endemism; however, previous authors have concluded that there is no direct coevolutionary relationship since these bees specialize on entire plant genera rather than individual species. Grasshoppers are generalist feeders, and their diversification may have more to do with low mobility and sexual selection than host plant use. California illustrates that considerable plant diversification can occur without triggering a similar radiation in animals, despite long-standing theory on coevolution and codiversification (e.g., Ehrlich and Raven 1964; Whittaker 1972).

Plants are not animals. Recent global analyses illustrate three notable facts about the five mediterranean climate regions. First, these regions are outstanding global hotspots of plant endemism; second, they are almost the only hotspots of plant endemism outside of the tropics; and third, they are the only plant endemism hotspots that are not also rich in vertebrate animal endemism. Together, this shows that there is something about either the contemporary mediterranean climate or its history that provokes high endemism in plants but not in animals. The traditional explanation would be that the short growing season and long summer drought impose special adaptive challenges for plants, and the onset of these conditions stimulated rapid speciation in certain pre-adapted plant lineages (a subtle variation on this idea would be that these climatic conditions selected for plant lineages that have high speciation rates wherever they are found). A related explanation (also predicting high speciation rates in plants) is that rugged topography and shifting climates created ideal conditions for localized allopatric divergence in plants while the greater mobility of animals prevented such an effect. New evidence seems more consistent with an alternative explanation: plants, as well as some moisture-sensitive and poorly dispersing animals, were protected from climatically caused extinctions in the relatively stable climates of the future mediterranean regions (also see section below).

Low extinction may be as significant as elevated speciation. In striking contrast to the conventional wisdom, recent phylogenetic analyses have shown that speciation rates are not higher, but rather extinction rates are lower, for plant lineages within the California Floristic Province compared to the same lineages elsewhere. This seems to contradict the idea of the Floristic Province as being rich in relatively young endemic species, but it can be reconciled if the equable climate permitted survival of the rapidly speciating lineages that gave rise to neoendemics. This result also agrees with global analyses finding that climatic stability is linked to high endemism and with regional analyses suggesting that the richest floras within the mediterranean regions are found where the climate was most stable during the Pleistocene. The equable climate has also permitted the survival of the paleoendemic component of both California's flora and its fauna. Over 100 plants can be interpreted as mesic (north-temperate) paleoendemics. The survival of mesic relictual taxa is also significant in certain groups of animals, notably those with dispersal that is either intrinsically low (primitive arachnids) or limited by the requirement for wet habitats (fish), or both (amphibians).

Biogeographic barriers may not play a central role in high Californian endemism. In Californian animals, there is abundant evidence linking genetic breaks in multiple groups to plate movements, the uplift of the mountain ranges, historic seaways, the Monterey and San Francisco Bays, and the aridification of the Central Valley (e.g., Calsbeek et al. 2003; Davis et al. 2008). In the most sedentary animals, such as the phalangodid harvestmen and some amphibians, these breaks have contributed to creating pairs or complexes of ecologically similar allopatric species. From a slightly larger perspective, these vicariant processes may not be strongly linked to Californian endemism, however. Rugged topography cannot explain why amphibians and fish are endemic-rich in California but nowhere else in the West. Conversely, rugged topography seems to have stimulated as much speciation in grasshoppers in Colorado as California (Knowles and Otte 2000). The five mediterranean regions are not collectively distinctive for their extreme topography, nor is endemism highest in the most geologically or topographically complex mediterranean regions. Mountain ranges throughout North America have generated phylogeographic breaks without generating high endemism, and California is not particularly outstanding as a center for phylogeographic breaks (Swenson and Howard 2005).

Physical heterogeneity also may not be the key to high endemism. Anyone familiar with the tremendous topographic and geologic heterogeneity of California's landscapes is likely to consider this as a prime explanation for its high biological diversity. This may be a lesser factor in its endemism, much of which is concentrated within the mediterranean climate and vegetation, even though a great deal of physical diversity is found in the high mountains and deserts. Likewise, there are complex landscapes in the Cascades, Rockies, and Great Basin ranges and elsewhere in the nonmediterranean parts of the temperate zone that do not support high levels of endemism. Several analyses have failed to find correlations between endemic plant diversity and environmental heterogeneity within California (Thorne, Viers et al. 2009; Kraft et al. 2010).

Diversity and composition of local ecological communities are shaped by historical and evolutionary processes. The biological coherence of the California Floristic Province makes it ideal for considering how large-scale evolutionary processes shape local ecological communities. Within the province, rainier climates have higher plant productivity, and these regions harbor the most diverse plant communities, a pattern that fits broad global trends, although it runs counter to the standard latitudinal diversity gradient. Ecologists would traditionally explain such a positive productivity-diversity relationship based on local processes such as resource availability and species interactions, but several pieces of evidence argue against this. First, the relationship is much stronger among species belonging to north-temperate than subtropical semiarid lineages; second, it is stronger at regional scales than local scales; and third, there is no evidence that the availability or partioning of niche space is correlated with local-scale plant diversity (Harrison and Grace 2007; Anacker and Harrison 2013). Some evolutionists, in contrast, might argue that the pattern results from a positive effect of productivity on speciation rates, but this does not fit the evidence either; the productivity-diversity pattern is no stronger (if anything, it is weaker) among neoendemics than other species. Instead, the relationship of climatic productivity to community diversity arises from the survival of north-temperate relicts in wetter environments, indicating that California's long-term climatic history has left a definitive imprint on its plant communities at the local as well as the regional scale.

Creative approaches are needed to slow the loss of biodiversity. With its hundreds of recognized rare species and its rapidly urbanizing natural habitats, California has long been at the forefront of clashes between development and the preservation of biodiversity. It has been a testing ground for alternative approaches that include conservation easements, conservation banking, advance mitigation, and large-scale regional planning. California's NCCP Act (as well as its late CALFED experiment) represents a hybrid approach to conservation: it uses the strong, prohibitive protections embodied by the Endangered Species Act to force competing parties to come to the table but at the same time encourages the development of solutions through collaboration and innovation (Babbitt 2005). Regional conservation planning is still very much a work in progress, but it offers one of the few hopes for setting aside meaningful networks of conservation land in the urbanizing parts of California that support so much of its biological diversity.

Preliminary List of Plant Species Endemic to the California Floristlc Province

Families and Full Species Only

AGAVACEAE

Camassia howellii, OR

Chlorogalum angustifolium, CA, OR

Chlorogalum grandiflorum, CA

Chlorogalum parviflorum, CA, Baja

Chlorogalum pomeridianum, CA, OR

Chlorogalum purpureum, CA

Hastingsia alba, CA, OR

Hastingsia bracteosa, OR

Hastingsia serpentinicola, CA, OR

ALISMATACEAE

Sagittaria sanfordii, CA

ALLIACEAE

Allium abramsii, CA

Allium bolanderi, CA, OR

Allium burlewii, CA

Allium cratericola, CA

Allium crispum, CA

Allium diabolense, CA

Allium dichlamydeum, CA

Allium eurotophilum, Baja

Allium falcifolium, CA, OR

Allium haematochiton, CA, Baja

Allium hickmanii, CA

Allium hoffmanii, CA

Allium howellii, CA

Allium hyalinum, CA

Allium jepsonii, CA

Allium marvinii, CA

Allium membranaceum, CA

Allium monticola, CA

Allium munzii, CA

Allium parryi, CA, Baja

Allium praecox, CA, Baja

Allium sanbornii, CA, OR

Allium serra, CA

Allium sharsmithiae, CA

Allium shevockii, CA

Allium siskiyouense, CA, OR

Allium tribracteatum, CA

Allium tuolumnense, CA

Allium unifolium, CA, OR

Allium yosemitense, CA

ANACARDIACEAE

Malosma laurina, CA, Baja

Rhus integrifolia, CA, Baja

APIACEAE

Angelica breweri, CA

Angelica californica, CA

Angelica callii, CA

Angelica tomentosa, CA, OR

Apiastrum angustifolium, CA, Baja

Eryngium aristulatum, CA, Baja

Eryngium armatum, CA

Eryngium castrense, CA

Eryngium constancei, CA

Eryngium jepsonii, CA

Eryngium pendletonense, CA

Eryngium pinnatisectum, CA

Eryngium racemosum, CA

Eryngium spinosepalum, CA

Eryngium vaseyi, CA

Ligusticum californicum, CA, OR

Lilaeopsis masonii, CA

Lomatium californicum, CA, OR

Lomatium caruifolium, CA

Lomatium ciliolatum, CA

Lomatium congdonii, CA

Lomatium cookii, OR

Lomatium dasycarpum, CA, Baja

Lomatium engelmannii, CA, OR

Lomatium hallii, CA, OR

APIACEAE Continued

Lomatium hooveri, CA

Lomatium howellii, CA, OR

Lomatium insulare, CA, Baja

Lomatium lucidum, CA, Baja

Lomatium macrocarpum, CA, OR

Lomatium marginatum, CA

Lomatium observatorium, CA

Lomatium parvifolium, CA

Lomatium peckianum, CA, OR

Lomatium repostum, CA

Lomatium shevockii, CA

Lomatium stebbinsii, CA

Lomatium torreyi, CA

Lomatium tracyi, CA

Oreonana clementis, CA

Oreonana purpurascens, CA

Oreonana vestita, CA

Perideridia bacigalupii, CA

Perideridia californica, CA

Perideridia howellii, CA, OR

Perideridia kelloggii, CA

Perideridia leptocarpa, CA

Perideridia pringlei, CA

Sanicula arguta, CA, Baja

Sanicula bipinnata, CA

Sanicula hoffmannii, CA

Sanicula laciniata, CA, OR

Sanicula maritima, CA

Sanicula moranii, Baja

Sanicula peckiana, CA, OR

Sanicula saxatilis, CA

Sanicula tracyi, CA

Sanicula tuberosa, CA, OR

Tauschia arguta, CA, Baja

Tauschia glauca, CA, OR

Tauschia hartwegii, CA

Tauschia howellii, CA, OR

Tauschia kelloggii, CA, OR

APOCYNACEAE

Asclepias solanoana, CA

ARECACEAE

Brahea edulis, Baja

ARISTOLOCHIACEAE

Aristolochia californica, CA

Asarum hartwegii, CA

Asarum lemmonii, CA

Asarum marmoratum, CA, OR

ASPLENIACEAE

Asplenium vespertinum, CA

ASTERACEAE

Achyrachaena mollis, CA, OR

Acourtia microcephala, CA, Baja

Adenothamnus validus, Baja

Ageratina shastensis, CA

Agoseris hirsuta, CA

Amblyopappus pusillus, CA, Baja

Ambrosia chenopodiifolia, CA, Baja

Ambrosia flexuosa, Baja

Ambrosia pumila, CA, Baja

Ancistrocarphus keilii, CA

Anisocarpus scabridus, CA

Antennaria suffrutescens, CA, OR

Arnica cernua, CA, OR

Arnica dealbata, CA

Arnica spathulata, CA, OR

Arnica venosa, CA

Arnica viscosa, CA

Artemisia californica, CA, Baja

Artemisia nesiotica, CA

Artemisia palmeri, CA, Baja

Baccharis malibuensis, CA

ASTERACEAE Continued

Baccharis plummerae, CA

Baccharis vanessae, CA

Baeriopsis guadalupensis, Baja

Balsamorhiza lanata, CA

Balsamorhiza sericea, CA, OR

Benitoa occidentalis, CA

Blennosperma bakeri, CA

Blennosperma nanum, CA

Blepharizonia laxa, CA

Blepharizonia plumosa, CA

Brickellia greenei, CA, OR

Brickellia subsessilis, Baja

Brickellia vollmeri, Baja

Calycadenia fremontii, CA, OR

Calycadenia hooveri, CA

Calycadenia micrantha, CA

Calycadenia mollis, CA

Calycadenia oppositifolia, CA

Calycadenia pauciflora, CA

Calycadenia spicata, CA

Calycadenia villosa, CA

Carlquistia muirii, CA

Centromadia fitchii, CA, OR

Centromadia parryi, CA

Centromadia perennis, Baja

Chaenactis artemisiifolia, CA, Baja

Chaenactis parishii, CA

Chaenactis santolinoides, CA

Chaenactis suffrutescens, CA

Cirsium andrewsii, CA

Cirsium ciliolatum, CA, OR

Cirsium crassicaule, CA

Cirsium fontinale, CA

Cirsium hydrophilum, CA

Cirsium praeteriens, CA

Cirsium quercetorum, CA

Cirsium rhothophilum, CA

Cirsium trachylomum, Baja

Constancea nevinii, CA

Deinandra bacigalupii, CA

Deinandra clementina, CA

Deinandra conjugens, CA

Deinandra corymbosa, CA

Deinandra fasciculata, CA, Baja

Deinandra floribunda, CA

Deinandra frutescens, Baja

Deinandra greeneana, Baja

Deinandra halliana, CA

Deinandra increscens, CA

Deinandra martirensis, Baja

Deinandra minthornii, CA

Deinandra pallida, CA

Deinandra palmeri, Baja

Deinandra paniculata, CA

Deinandra pentactis, CA

Deinandra peninsularis, Baja

Deinandra streetsii, Baja

Eastwoodia elegans, CA

Ericameria arborescens, CA

Ericameria brachylepis, CA, Baja

Ericameria ericoides, CA, Baja

Ericameria fasciculata, CA

Ericameria juarezensis, Baja

Ericameria martirensis, Baja

Ericameria ophitidis, CA

Ericameria parishii, CA, Baja

Erigeron aequifolius, CA

Erigeron barbellulatus, CA

Erigeron biolettii, CA

Erigeron blochmaniae, CA

Erigeron cervinus, CA, OR

Erigeron elmeri, CA

Erigeron greenei, CA

ASTERACEAE Continued

Erigeron klamathensis, CA, OR

Erigeron lassenianus, CA

Erigeron maniopotamicus, CA

Erigeron mariposanus, CA

Erigeron miser, CA

Erigeron multiceps, CA

Erigeron parishii, CA

Erigeron reductus, CA

Erigeron robustior, CA

Erigeron sanctarum, CA

Erigeron serpentinus, CA

Erigeron stanselliae, OR

Erigeron supplex, CA

Eriophyllum congdonii, CA

Eriophyllum jepsonii, CA

Eriophyllum latilobum, CA

Eriophyllum nubigenum, CA

Eucephalus glabratus, CA, OR

Eucephalus tomentellus, CA, OR

Eucephalus vialis, CA

Geraea viscida, CA

Grindelia hallii, CA

Gutierrezia californica, CA, Baja

Harmonia doris-nilesiae, CA

Harmonia guggolziorum, CA

Harmonia hallii, CA

Harmonia nutans, CA

Harmonia stebbinsii, CA

Hazardia berberidis, Baja

Hazardia cana, CA

Hazardia detonsa, CA

Hazardia enormidens, Baja

Hazardia ferrisiae, Baja

Hazardia orcuttii, CA, Baja

Hazardia rosarica, Baja

Hazardia squarrosa, CA, Baja

Hazardia stenolepis, CA

Hazardia vernicosa, Baja

Hazardia whitneyi, CA, OR

Helenium puberulum, CA, Baja

Helianthella castanea, CA

Helianthus bolanderi, CA, OR

Helianthus californicus, CA

Helianthus exilis, CA

Helianthus gracilentus, CA, Baja

Helianthus inexpectatus, CA

Hemizonia congesta, CA, OR

Hesperevax acaulis, CA, OR

Hesperevax caulescens, CA

Heterotheca brandegeei, Baja

Heterotheca monarchensis, CA

Heterotheca shevockii, CA

Hieracium argutum, CA

Holocarpha heermannii, CA

Holocarpha macradenia, CA

Holocarpha obconica, CA

Holocarpha virgata, CA

Hulsea brevifolia, CA

Hulsea californica, CA

Hulsea mexicana, CA, Baja

Isocoma arguta, CA

Iva hayesiana, CA, Baja

Jensia rammii, CA

Jensia yosemitana, CA

Lagophylla dichotoma, CA

Lagophylla glandulosa, CA

Lagophylla minor, CA

Lasthenia burkei, CA

Lasthenia chrysantha, CA

Lasthenia conjugens, CA

Lasthenia ferrisiae, CA

Lasthenia fremontii, CA

Lasthenia leptalea, CA

ASTERACEAE Continued

Lasthenia minor, CA

Lasthenia platycarpha, CA

Layia carnosa, CA

Layia chrysanthemoides, CA

Layia discoidea, CA

Layia fremontii, CA

Layia gaillardioides, CA

Layia heterotricha, CA

Layia hieracioides, CA

Layia jonesii, CA

Layia leucopappa, CA

Layia munzii, CA

Layia pentachaeta, CA

Layia septentrionalis, CA

Leptosyne douglasii, CA

Leptosyne gigantea, CA

Leptosyne hamiltonii, CA

Leptosyne maritima, CA, Baja

Leptosyne stillmanii, CA

Lessingia arachnoidea, CA

Lessingia germanorum, CA

Lessingia hololeuca, CA

Lessingia leptoclada, CA

Lessingia micradenia, CA

Lessingia nana, CA

Lessingia nemaclada, CA

Lessingia pectinata, CA

Lessingia ramulosa, CA

Lessingia tenuis, CA

Lessingia virgata, CA

Madia anomala, CA

Madia radiata, CA

Madia subspicata, CA

Malacothrix foliosa, CA

Malacothrix incana, CA

Malacothrix indecora, CA

Malacothrix insularis, Baja

Malacothrix junakii, CA

Malacothrix phaeocarpa, CA

Malacothrix saxatilis, CA

Malacothrix similis, CA

Malacothrix squalida, CA

Micropus amphibolus, CA

Microseris acuminata, CA, OR

Microseris campestris, CA

Microseris douglasii, CA, OR, Baja

Microseris elegans, CA, Baja

Microseris howellii, OR

Microseris paludosa, CA

Monolopia congdonii, CA

Monolopia gracilens, CA

Monolopia major, CA

Monolopia stricta, CA

Munzothamnus blairii, CA

Oreostemma elatum, CA

Oreostemma peirsonii, CA

Orochaenactis thysanocarpha, CA

Osmadenia tenella, CA

Packera bernardina, CA

Packera breweri, CA

Packera clevelandii, CA

Packera ganderi, CA

Packera greenei, CA

Packera hesperia, CA, OR

Packera ionophylla, CA

Packera layneae, CA

Packera pauciflora, CA

Pentachaeta alsinoides, CA

Pentachaeta aurea, CA

Pentachaeta bellidiflora, CA

Pentachaeta exilis, CA

Pentachaeta fragilis, CA

Pentachaeta lyonii, CA

ASTERACEAE Continued

Perityle incana, Baja

Phalacroseris bolanderi, CA

Pseudobahia bahiifolia, CA

Pseudobahia heermannii, CA

Pseudobahia peirsonii, CA

Pseudognaphalium beneolens, CA, Baja

Pseudognaphalium biolettii, CA, Baja

Pseudognaphalium microcephalum, CA, Baja

Pseudognaphalium ramosissimum, CA

Pseudognaphalium roseum, CA

Pyrrocoma lucida, CA

Raillardella pringlei, CA

Raillardella scaposa, CA

Rudbeckia californica, CA, OR

Rudbeckia klamathensis, CA

Saussurea americana, CA, OR

Senecio aphanactis, CA

Senecio astephanus, CA

Senecio blochmaniae, CA

Senecio cedrosensis, Baja

Senecio clarkianus, CA

Senecio lyonii, CA, Baja

Senecio martirensis, Baja

Senecio palmeri, Baja

Solidago guiradonis, CA

Solidago spathulata, CA

Sphaeromeria martirensis, Baja

Stebbinsoseris decipiens, CA

Stebbinsoseris heterocarpa, CA

Stenotus pulvinatus, Baja

Stephanomeria cichoriacea, CA

Stephanomeria diegensis, CA, Baja

Stephanomeria elata, CA, OR

Stephanomeria guadalupensis, Baja

Stephanomeria monocephala, Baja

Stylocline citroleum, CA

Stylocline masonii, CA

Symphyotrichum defoliatum, CA

Symphyotrichum greatae, CA

Symphyotrichum hendersonii, CA, OR

Symphyotrichum lentum, CA

Syntrichopappus lemmonii, CA

Taraxacum californicum, CA

Tetradymia comosa, CA

Tonestus eximius, CA

Tonestus lyallii, CA

Tracyina rostrata, CA

Trichocoronis wrightii, CA

Venegasia carpesioides, CA, Baja

Verbesina dissita, CA, Baja

Verbesina hastata, Baja

Viguiera purisimae, CA, Baja

Wyethia bolanderi, CA

Wyethia elata, CA

Wyethia glabra, CA

Wyethia helenioides, CA

Wyethia invenusta, CA

Wyethia longicaulis, CA

Wyethia ovata, CA

Wyethia reticulata, CA

Xanthisma junceum, CA, Baja

Xanthisma wigginsii, Baja

BATACEAE

Batis maritima, CA, Baja

BERBERIDACEAE

Achlys californica, CA, OR

Berberis claireae, Baja

Berberis higginsiae, CA, Baja

Berberis nevinii, CA

BORAGINACEAE

Amsinckia douglasiana, CA

Amsinckia eastwoodiae, CA

BORAGINACEAE Continued

Amsinckia furcata, CA

Amsinckia grandiflora, CA

Amsinckia inepta, Baja

Amsinckia lunaris, CA

Amsinckia spectabilis, CA, Baja

Cryptantha clevelandii, CA, Baja

Cryptantha corollata, CA

Cryptantha crinita, CA

Cryptantha crymophila, CA

Cryptantha dissita, CA

Cryptantha excavata, CA

Cryptantha foliosa, Baja

Cryptantha hispidula, CA, OR

Cryptantha hooveri, CA

Cryptantha incana, CA

Cryptantha mariposae, CA

Cryptantha micromeres, CA

Cryptantha microstachys, CA

Cryptantha nemaclada, CA

Cryptantha rattanii, CA

Cryptantha sparsiflora, CA

Cryptantha spithamaea, CA

Cryptantha traskiae, CA

Draperia systyla, CA

Eriodictyon altissimum, CA

Eriodictyon capitatum, CA

Eriodictyon crassifolium, CA

Eriodictyon sessilifolium, Baja

Eriodictyon tomentosum, CA

Eriodictyon traskiae, CA

Hackelia amethystina, CA

Hackelia mundula, CA

Hackelia nervosa, CA

Hackelia sharsmithii, CA

Hackelia velutina, CA

Mertensia bella, CA, OR

Nemophila maculata, CA, OR

Nemophila pulchella, CA

Phacelia brachyloba, CA, Baja

Phacelia breweri, CA

Phacelia californica, CA

Phacelia ciliata, CA

Phacelia congdonii, CA

Phacelia cookei, CA

Phacelia corymbosa, CA, OR

Phacelia dalesiana, CA

Phacelia davidsonii, CA

Phacelia divaricata, CA

Phacelia egena, CA, OR

Phacelia eisenii, CA

Phacelia exilis, CA

Phacelia floribunda, CA, Baja

Phacelia grandiflora, CA, Baja

Phacelia greenei, CA

Phacelia grisea, CA

Phacelia hubbyi, CA

Phacelia hydrophylloides, CA

Phacelia imbricata, CA, Baja

Phacelia insularis, CA

Phacelia ixodes, Baja

Phacelia keckii, CA

Phacelia leonis, CA, OR

Phacelia lyonii, CA

Phacelia malvifolia, CA, OR

Phacelia marcescens, CA

Phacelia mohavensis, CA

Phacelia orogenes, CA

Phacelia peckii, OR

Phacelia phacelioides, CA

Phacelia phyllomanica, Baja

Phacelia platyloba, CA

Phacelia procera, CA, OR

Phacelia quickii, CA

BORAGINACEAE Continued

Phacelia racemosa, CA

Phacelia rattanii, CA, OR

Phacelia stebbinsii, CA

Phacelia stellaris, CA

Phacelia suaveolens, CA

Phacelia umbrosa, CA

Phacelia vallicola, CA

Plagiobothrys acanthocarpus, CA, Baja

Plagiobothrys austiniae, CA, OR

Plagiobothrys chorisianus, CA

Plagiobothrys diffusus, CA

Plagiobothrys distantiflorus, CA

Plagiobothrys fulvus, CA, OR

Plagiobothrys glaber, CA

Plagiobothrys humistratus, CA

Plagiobothrys hystriculus, CA

Plagiobothrys infectivus, CA

Plagiobothrys lamprocarpus, OR

Plagiobothrys lithocaryus, CA

Plagiobothrys reticulatus, CA, OR

Plagiobothrys scriptus, CA

Plagiobothrys strictus, CA

Plagiobothrys torreyi, CA

Plagiobothrys trachycarpus, CA

Plagiobothrys uncinatus, CA

Plagiobothrys undulatus, CA

BRASSICACEAE

Arabis aculeolata, OR

Arabis blepharophylla, CA

Arabis mcdonaldiana, CA, OR

Arabis modesta, CA, OR

Athysanus unilateralis, CA, OR, Baja

Boechera arcuata, CA

Boechera breweri, CA, OR

Boechera californica, CA, Baja

Boechera constancei, CA

Boechera evadens, CA

Boechera hoffmannii, CA

Boechera johnstonii, CA

Boechera parishii, CA

Boechera peirsonii, CA

Boechera pygmaea, CA

Boechera repanda, CA

Boechera rigidissima, CA

Boechera rollei, CA

Boechera rubicundula, CA

Boechera serpenticola, CA

Boechera shevockii, CA

Boechera tiehmii, CA

Boechera tularensis, CA

Boechera ultraalsa, CA

Cardamine bellidifolia, CA

Cardamine pachystigma, CA

Caulanthus amplexicaulis, CA

Caulanthus anceps, CA

Caulanthus flavescens, CA

Caulanthus heterophyllus, CA, Baja

Caulanthus lemmonii, CA

Dithyrea maritima, CA

Draba asterophora, CA

Draba aureola, CA

Draba carnosula, CA

Draba corrugata, CA, Baja

Draba cruciata, CA

Draba demareei, Baja

Draba howellii, CA, OR

Draba lemmonii, CA

Draba longisquamosa, CA

Draba pterosperma, CA

Draba saxosa, CA

Draba sharsmithii, CA

Draba sierrae, CA

Erysimum ammophilum, CA

BRASSICACEAE Continued

Erysimum franciscanum, CA

Erysimum insulare, CA

Erysimum menziesii, CA

Erysimum moranii, Baja

Erysimum suffrutescens, CA

Erysimum teretifolium, CA

Lepidium jaredii, CA

Lepidium latipes, CA, Baja

Lepidium oxycarpum, CA

Lepidium strictum, CA, OR

Nasturtium gambelii, CA

Physaria palmeri, Baja

Physaria peninsularis, Baja

Rorippa sphaerocarpa, CA

Rorippa subumbellata, CA

Rorippa tenerrima, CA, Baja

Sibara filifolia, CA

Sibaropsis hammittii, CA

Smelowskia ovalis, CA

Streptanthus barbatus, CA

Streptanthus barbiger, CA

Streptanthus batrachopus, CA

Streptanthus bernardinus, CA

Streptanthus brachiatus, CA

Streptanthus breweri, CA

Streptanthus callistus, CA

Streptanthus campestris, CA, Baja

Streptanthus diversifolius, CA

Streptanthus drepanoides, CA

Streptanthus farnsworthianus, CA

Streptanthus fenestratus, CA

Streptanthus glandulosus, CA, OR

Streptanthus gracilis, CA

Streptanthus hesperidis, CA

Streptanthus hispidus, CA

Streptanthus howellii, CA, OR

Streptanthus insignis, CA

Streptanthus longisiliquus, CA

Streptanthus morrisonii, CA

Streptanthus oblanceolatus, CA

Streptanthus polygaloides, CA

Streptanthus tortuosus, CA, OR

Streptanthus vernalis, CA

Streptanthus vimineus, CA

Subularia aquatica, CA

Thelypodium stenopetalum, CA

Thysanocarpus conchuliferus, CA

Thysanocarpus radians, CA, OR

Tropidocarpum californicum, CA

Tropidocarpum capparideum, CA

CACTACEAE

Bergerocactus emoryi, CA, Baja

Cochemiea pondii, Baja

Cylindropuntia prolifera, CA, Baja

Echinocereus mombergerianus, Baja

Echinocereus pacificus, Baja

Ferocactus viridescens, CA, Baja

Mammillaria louisae, Baja

Mammillaria neopalmeri, Baja

Opuntia fragilis, CA

Opuntia littoralis, CA, Baja

Opuntia oricola, CA

CALYCANTHACEAE

Calycanthus occidentalis, CA, OR

CAMPANULACEAE

Campanula angustiflora, CA

Campanula californica, CA

Campanula exigua, CA

Campanula griffinii, CA

Campanula rotundifolia, CA, OR

Campanula sharsmithiae, CA

CAMPANULACEAE Continued

Campanula shetleri, CA

Campanula wilkinsiana, CA

Downingia bella, CA

Downingia concolor, CA

Downingia montana, CA

Downingia ornatissima, CA

Downingia pulchella, CA

Downingia pusilla, CA

Githopsis diffusa, CA

Githopsis pulchella, CA

Githopsis tenella, CA

Legenere limosa, CA

Lobelia dunnii, CA, Baja

Nemacladus calcaratus, CA

Nemacladus interior, CA

Nemacladus montanus, CA

Nemacladus secundiflorus, CA

Nemacladus twisselmannii, CA

CAPRIFOLIACEAE

Lonicera cauriana, CA

Lonicera subspicata, CA, Baja

CARYOPHYLLACEAE

Arenaria lanuginosa, CA

Arenaria paludicola, CA

Eremogone cliftonii, CA

Eremogone ursina, CA

Minuartia decumbens, CA

Minuartia howellii, CA, OR

Minuartia obtusiloba, CA

Minuartia rosei, CA

Minuartia stolonifera, CA

Paronychia ahartii, CA

Polycarpon depressum, CA

Pseudostellaria sierrae, CA

Silene aperta, CA

Silene bolanderi, CA, OR

Silene bridgesii, CA, OR

Silene grayi, CA, OR

Silene invisa, CA

Silene laciniata, CA, Baja

Silene marmorensis, CA

Silene parishii, CA

Silene salmonacea, CA

Silene scouleri, CA, OR

Silene serpentinicola, CA

Silene suksdorfii, CA

Spergularia canadensis, CA

Stellaria littoralis, CA

CHENOPODIACEAE

Aphanisma blitoides, CA, Baja

Atriplex californica, CA

Atriplex cordulata, CA

Atriplex coronata, CA

Atriplex coulteri, CA

Atriplex depressa, CA

Atriplex joaquinana, CA

Atriplex leucophylla, CA

Atriplex minuscula, CA

Atriplex pacifica, CA, Baja

Atriplex persistens, CA

Atriplex subtilis, CA

Atriplex tularensis, CA

Atriplex watsonii, CA, Baja

Chenopodium flabellifolium, Baja

Chenopodium littoreum, CA

Salicornia bigelovii, CA, Baja

Suaeda californica, CA, Baja

Suaeda esteroa, CA, Baja

Suaeda taxifolia, CA

CISTACEAE

Helianthemum greenei, CA

Helianthemum scoparium, CA, Baja

CONVOLVULACEAE

Calystegia collina, CA

Calystegia macrostegia, CA, Baja

Calystegia malacophylla, CA

Calystegia purpurata, CA

Calystegia stebbinsii, CA

Calystegia subacaulis, CA

Convolvulus simulans, CA

Cuscuta brachycalyx, CA

Cuscuta jepsonii, CA

Cuscuta obtusiflora, CA

Dichondra occidentalis, CA

CORNACEAE

Cornus sessilis, CA

CRASSULACEAE

Crassula aquatica, CA

Crassula solieri, CA

Dudleya anomala, Baja

Dudleya anthonyi, Baja

Dudleya attenuata, CA, Baja

Dudleya blochmaniae, CA

Dudleya brevifolia, CA

Dudleya brittonii, Baja

Dudleya caespitosa, CA

Dudleya campanulata, Baja

Dudleya candelabrum, CA

Dudleya candida, Baja

Dudleya cymosa, CA

Dudleya densiflora, CA

Dudleya edulis, CA, Baja

Dudleya formosa, Baja

Dudleya gnoma, CA

Dudleya greenei, CA

Dudleya guadalupensis, Baja

Dudleya linearis, Baja

Dudleya multicaulis, CA

Dudleya nesiotica, CA

Dudleya pachyphytum, Baja

Dudleya palmeri, CA

Dudleya parva, CA

Dudleya pauciflora, Baja

Dudleya pulverulenta, CA, Baja

Dudleya stolonifera, CA

Dudleya traskiae, CA

Dudleya variegata, CA, Baja

Dudleya verityi, CA

Dudleya virens, CA, Baja

Dudleya viscida, CA

Sedella leiocarpa, CA

Sedella pentandra, CA

Sedella pumila, CA

Sedum albomarginatum, CA

Sedum lanceolatum, CA, OR

Sedum moranii, OR

Sedum oblanceolatum, CA, OR

Sedum radiatum, CA, OR

CROSSOSOMATACEAE

Crossosoma californicum, CA

CUCURBITACEAE

Marah guadalupensis, Baja

Marah horrida, CA

Marah watsonii, CA

CUPRESSACEAE

Hesperocyparis abramsiana, CA

Hesperocyparis forbesii, CA, Baja

Hesperocyparis goveniana, CA

Hesperocyparis guadalupensis, Baja

Hesperocyparis macnabiana, CA, OR

Hesperocyparis macrocarpa, CA

Hesperocyparis montana, Baja

Hesperocyparis nevadensis, CA

Hesperocyparis pygmaea, CA

CUPRESSACEAE Continued

Hesperocyparis sargentii, CA, OR

Hesperocyparis stephensonii, CA, Baja

Sequoia sempervirens, CA, OR

Sequoiadendron giganteum, CA

CYPERACEAE

Bolboschoenus robustus, CA, Baja

Bulbostylis capillaris, CA, OR

Carex bolanderi, CA, OR

Carex congdonii, CA

Carex davyi, CA

Carex fissuricola, CA, OR

Carex globosa, CA, Baja

Carex gracilior, CA

Carex halliana, CA

Carex harfordii, CA

Carex hirtissima, CA

Carex klamathensis, CA, OR

Carex lasiocarpa, CA

Carex lemmonii, CA

Carex luzulifolia, CA

Carex mariposana, CA

Carex nervina, CA, OR

Carex nigricans, CA

Carex obispoensis, CA

Carex pansa, CA

Carex praeceptorum, CA

Carex proposita, CA

Carex saliniformis, CA

Carex sartwelliana, CA

Carex scabriuscula, CA, OR

Carex schottii, CA

Carex scoparia, CA

Carex senta, CA, Baja

Carex specifica, CA

Carex spectabilis, CA, OR

Carex subbracteata, CA, OR

Carex tiogana, CA

Carex tompkinsii, CA

Carex triquetra, CA, Baja

Carex viridula, CA, OR

Cyperus acuminatus, CA, OR

Dulichium arundinaceum, CA

Eleocharis atropurpurea, CA

Eleocharis bernardina, CA

Eleocharis decumbens, CA, OR

Eleocharis parvula, CA

Eleocharis radicans, CA

Eleocharis torticulmis, CA

Isolepis carinata, CA, Baja

Lipocarpha aristulata, CA

Lipocarpha micrantha, CA

Lipocarpha occidentalis, CA

Rhynchospora californica, CA

Rhynchospora capitellata, CA, OR

Rhynchospora globularis, CA

Schoenoplectus saximontanus, CA

Schoenoplectus subterminalis, CA, OR

Scirpus diffusus, CA

Trichophorum clementis, CA

DRYOPTERIDACEAE

Polystichum dudleyi, CA

Polystichum kruckebergii, CA

Polystichum lemmonii, CA, OR

ELATINACEAE

Bergia texana, CA

Elatine californica, CA, Baja

Elatine heterandra, CA

ERICACEAE

Arctostaphylos andersonii, CA

Arctostaphylos auriculata, CA

Arctostaphylos australis, Baja

Arctostaphylos bakeri, CA

ERICACEAE Continued

Arctostaphylos bolensis, Baja

Arctostaphylos canescens, CA, OR

Arctostaphylos catalinae, CA

Arctostaphylos confertiflora, CA

Arctostaphylos crustacea, CA

Arctostaphylos cruzensis, CA

Arctostaphylos densiflora, CA

Arctostaphylos edmundsii, CA

Arctostaphylos franciscana, CA

Arctostaphylos gabilanensis, CA

Arctostaphylos glandulosa, CA, OR, Baja

Arctostaphylos glutinosa, CA

Arctostaphylos hispidula, CA, OR

Arctostaphylos hookeri, CA

Arctostaphylos hooveri, CA

Arctostaphylos imbricata, CA

Arctostaphylos incognita, Baja

Arctostaphylos insularis, CA

Arctostaphylos klamathensis, CA

Arctostaphylos luciana, CA

Arctostaphylos malloryi, CA

Arctostaphylos manzanita, CA

Arctostaphylos mewukka, CA

Arctostaphylos montana, CA

Arctostaphylos montaraensis, CA

Arctostaphylos montereyensis, CA

Arctostaphylos morroensis, CA

Arctostaphylos myrtifolia, CA

Arctostaphylos nissenana, CA

Arctostaphylos nortensis, CA, OR

Arctostaphylos nummularia, CA

Arctostaphylos obispoensis, CA

Arctostaphylos ohloneana, CA

Arctostaphylos osoensis, CA

Arctostaphylos otayensis, CA

Arctostaphylos pacifica, CA

Arctostaphylos pajaroensis, CA

Arctostaphylos pallida, CA

Arctostaphylos parryana, CA

Arctostaphylos pechoensis, CA

Arctostaphylos pilosula, CA

Arctostaphylos pringlei, CA

Arctostaphylos pumila, CA

Arctostaphylos purissima, CA

Arctostaphylos rainbowensis, CA

Arctostaphylos refugioensis, CA

Arctostaphylos regismontana, CA

Arctostaphylos rudis, CA

Arctostaphylos sensitiva, CA

Arctostaphylos silvicola, CA

Arctostaphylos stanfordiana, CA

Arctostaphylos tomentosa, CA

Arctostaphylos virgata, CA

Arctostaphylos viridissima, CA

Arctostaphylos viscida, CA, OR

Cassiope mertensiana, CA

Comarostaphylis diversifolia, CA, Baja

Kalmiopsis leachiana, OR

Ornithostaphylos oppositifolia, CA, Baja

Phyllodoce breweri, CA

EUPHORBIACEAE

Chamaesyce hooveri, CA

Euphorbia benedicta, Baja

FABACEAE

Acmispon argophyllus, CA, Baja

Acmispon cytisoides, CA

Acmispon dendroideus, CA

Acmispon distichus, Baja

Acmispon flexuosus, Baja

Acmispon grandiflorus, CA, Baja

Acmispon junceus, CA

Acmispon prostratus, CA, Baja

FABACEAE Continued

Acmispon rubriflorus, CA

Amorpha apiculata, Baja

Astragalus agnicidus, CA

Astragalus anemophilus, Baja

Astragalus asymmetricus, CA

Astragalus austiniae, CA

Astragalus bicristatus, CA

Astragalus bolanderi, CA

Astragalus brauntonii, CA

Astragalus breweri, CA

Astragalus circumdatus, Baja

Astragalus claranus, CA

Astragalus clevelandii, CA

Astragalus congdonii, CA

Astragalus curtipes, CA

Astragalus deanei, CA

Astragalus ertterae, CA

Astragalus gambelianus, CA, OR

Astragalus harbisonii, Baja

Astragalus lentiformis, CA

Astragalus leucolobus, CA

Astragalus macrodon, CA

Astragalus miguelensis, CA

Astragalus morani, Baja

Astragalus nevinii, CA

Astragalus nuttallii, CA

Astragalus oocarpus, CA

Astragalus oxyphysus, CA

Astragalus oxyphysopsis, Baja

Astragalus pauperculus, CA

Astragalus pycnostachyus, CA

Astragalus rattanii, CA

Astragalus ravenii, CA

Astragalus sanctorum, Baja

Astragalus shevockii, CA

Astragalus subvestitus, CA

Astragalus tener, CA

Astragalus traskiae, CA

Hoita macrostachya, CA

Hoita orbicularis, CA

Hoita strobilina, CA

Hosackia incana, CA

Hosackia stipularis, CA

Hosackia yollabolliensis, CA

Lathyrus biflorus, CA

Lathyrus glandulosus, CA

Lathyrus jepsonii, CA

Lathyrus palustris, CA, OR

Lathyrus splendens, CA, Baja

Lupinus adsurgens, CA

Lupinus antoninus, CA

Lupinus apertus, CA

Lupinus benthamii, CA

Lupinus cervinus, CA

Lupinus chamissonis, CA

Lupinus citrinus, CA

Lupinus constancei, CA

Lupinus covillei, CA

Lupinus croceus, CA

Lupinus dalesiae, CA

Lupinus elatus, CA

Lupinus elmeri, CA

Lupinus fulcratus, CA

Lupinus gracilentus, CA

Lupinus grayi, CA

Lupinus guadalupensis, CA

Lupinus hirsutissimus, CA

Lupinus hyacinthinus, CA, Baja

Lupinus lapidicola, CA

Lupinus longifolius, CA, Baja

Lupinus ludovicianus, CA

Lupinus milo-bakeri, CA

Lupinus nanus, CA

FABACEAE Continued

Lupinus nipomensis, CA

Lupinus niveus, Baja

Lupinus obtusilobus, CA

Lupinus pachylobus, CA

Lupinus peirsonii, CA

Lupinus sericatus, CA

Lupinus spectabilis, CA

Lupinus stiversii, CA

Lupinus succulentus, CA, Baja

Lupinus tidestromii, CA

Lupinus tracyi, CA, OR

Lupinus truncatus, CA, Baja

Lupinus variicolor, CA

Pediomelum californicum, CA

Pickeringia montana, CA

Rupertia hallii, CA

Rupertia rigida, CA

Sophora leachiana, OR

Syrmatium watsonii, Baja

Thermopsis robusta, CA

Trifolium amoenum, CA

Trifolium bolanderi, CA

Trifolium breweri, CA, OR

Trifolium buckwestiorum, CA

Trifolium fucatum, CA, OR

Trifolium grayi, CA

Trifolium hydrophilum, CA

Trifolium jokerstii, CA

Trifolium lemmonii, CA

Trifolium olivaceum, CA

Trifolium palmeri, CA, Baja

Trifolium polyodon, CA

Trifolium trichocalyx, CA

Trifolium wigginsii, Baja

Vicia gigantea, CA

Vicia hassei, CA, OR

FAGACEAE

Quercus agrifolia, CA, Baja

Quercus berberidifolia, CA, Baja

Quercus dumosa, CA, Baja

Quercus durata, CA

Quercus engelmannii, CA, Baja

Quercus lobata, CA

Quercus pacifica, CA

Quercus parvula, CA

Quercus sadleriana, CA, OR

Quercus tomentella, CA, Baja

GARRYACEAE

Garrya buxifolia, CA, OR

Garrya congdonii, CA

Garrya veatchii, CA, Baja

GENTIANACEAE

Frasera neglecta, CA

Frasera tubulosa, CA

Gentiana fremontii, CA

Gentiana plurisetosa, CA, OR

Swertia perennis, CA

Zeltnera davyi, CA

Zeltnera trichantha, CA

GERANIACEAE

California macrophylla, CA, OR

Geranium bicknellii, CA, OR

GROSSULARIACEAE

Ribes amarum, CA

Ribes californicum, CA

Ribes canthariforme, CA

Ribes indecorum, CA, Baja

Ribes lasianthum, CA

Ribes malvaceum, CA, Baja

Ribes marshallii, CA, OR

Ribes sericeum, CA

GROSSULARIACEAE Continued
Ribes speciosum, CA, Baja
Ribes thacherianum, CA
Ribes tularense, CA
Ribes viburnifolium, CA, Baja
Ribes victoris, CA

HALORAGACEAE
Myriophyllum hippuroides, CA, OR

HYDRANGEACEAE
Carpenteria californica, CA

HYDROCHARITACEAE
Najas flexilis, CA

HYPERICACEAE
Hypericum concinnum, CA

IRIDACEAE
Iris bracteata, CA, OR
Iris fernaldii, CA
Iris hartwegii, CA
Iris longipetala, CA
Iris macrosiphon, CA
Iris munzii, CA
Iris purdyi, CA
Iris tenuissima, CA
Iris thompsonii, CA, OR
Sisyrinchium elmeri, CA

ISOETACEAE
Isoetes echinospora, CA, OR
Isoetes orcuttii, CA, OR, Baja

JUGLANDACEAE
Juglans californica, CA
Juglans hindsii, CA

JUNCACEAE
Juncus capillaris, CA
Juncus chlorocephalus, CA

Juncus digitatus, CA
Juncus drummondii, CA
Juncus duranii, CA
Juncus falcatus, CA
Juncus kelloggii, CA, OR
Juncus phaeocephalus, CA
Juncus textilis, CA
Juncus triformis, CA
Luzula divaricata, CA
Luzula orestera, CA
Luzula subcongesta, CA, OR

JUNCAGINACEAE
Triglochin palustris, CA
Triglochin striata, CA, OR

LAMIACEAE
Acanthomintha duttonii, CA
Acanthomintha ilicifolia, CA, Baja
Acanthomintha lanceolata, CA
Acanthomintha obovata, CA
Clinopodium chandleri, CA, Baja
Clinopodium ganderi, Baja
Clinopodium mimuloides, CA
Clinopodium palmeri, Baja
Hedeoma martirensis, Baja
Hedeoma matomiana, Baja
Lepechinia calycina, CA
Lepechinia cardiophylla, CA, Baja
Lepechinia fragrans, CA
Lepechinia ganderi, CA
Lepechinia rossii, CA
Monardella australis, CA, Baja
Monardella beneolens, CA
Monardella candicans, CA
Monardella douglasii, CA
Monardella follettii, CA
Monardella hypoleuca, CA

LAMIACEAE Continued

Monardella leucocephala, CA

Monardella macrantha, CA, Baja

Monardella nana, CA, Baja

Monardella palmeri, CA

Monardella pringlei, CA

Monardella saxicola, CA

Monardella sinuata, CA

Monardella stebbinsii, CA

Monardella stoneana, CA, Baja

Monardella thymifolia, Baja

Monardella undulata, CA

Monardella venosa, CA

Monardella viminea, CA

Monardella viridis, CA

Pogogyne abramsii, CA

Pogogyne clareana, CA

Pogogyne douglasii, CA

Pogogyne nudiuscula, CA

Pogogyne serpylloides, CA

Pogogyne zizyphoroides, CA, OR

Salvia brandegeei, CA, Baja

Salvia clevelandii, CA

Salvia leucophylla, CA

Salvia mellifera, CA

Salvia sonomensis, CA, OR

Salvia spathacea, CA

Scutellaria californica, CA

Scutellaria siphocampyloides, CA

Stachys bergii, CA

Stachys bullata, CA

Stachys chamissonis, CA

Stachys pycnantha, CA

Stachys stebbinsii, CA

Stachys stricta, CA

Trichostema lanatum, CA

Trichostema micranthum, CA

Trichostema ovatum, CA

Trichostema parishii, CA, Baja

Trichostema rubisepalum, CA

Trichostema ruygtii, CA

LILIACEAE

Calochortus albus, CA

Calochortus amabilis, CA

Calochortus amoenus, CA

Calochortus argillosus, CA

Calochortus catalinae, CA

Calochortus clavatus, CA

Calochortus coeruleus, CA

Calochortus concolor, CA, Baja

Calochortus dunnii, CA

Calochortus fimbriatus, CA, Baja

Calochortus howellii, OR

Calochortus indecorus, OR

Calochortus luteus, CA

Calochortus minimus, CA

Calochortus monanthus, CA

Calochortus nudus, CA, OR

Calochortus obispoensis, CA

Calochortus palmeri, CA

Calochortus persistens, CA

Calochortus plummerae, CA

Calochortus pulchellus, CA

Calochortus raichei, CA

Calochortus simulans, CA

Calochortus splendens, CA, Baja

Calochortus superbus, CA

Calochortus syntrophus, CA

Calochortus tiburonensis, CA

Calochortus umbellatus, CA

Calochortus venustus, CA

Calochortus vestae, CA

Calochortus weedii, CA, Baja

Calochortus westonii, CA

LILIACEAE Continued

Clintonia andrewsiana, CA, OR

Erythronium californicum, CA

Erythronium helenae, CA

Erythronium multiscapideum, CA

Erythronium pluriflorum, CA

Erythronium purpurascens, CA

Erythronium pusaterii, CA

Erythronium taylorii, CA

Erythronium tuolumnense, CA

Fritillaria agrestis, CA

Fritillaria biflora, CA, Baja

Fritillaria brandegeei, CA

Fritillaria eastwoodiae, CA, OR

Fritillaria falcata, CA

Fritillaria gentneri, CA, OR

Fritillaria liliacea, CA

Fritillaria micrantha, CA

Fritillaria ojaiensis, CA

Fritillaria pinetorum, CA

Fritillaria pluriflora, CA

Fritillaria purdyi, CA, OR

Fritillaria striata, CA

Fritillaria viridea, CA

Lilium bolanderi, CA, OR

Lilium humboldtii, CA

Lilium kelleyanum, CA

Lilium kelloggii, CA, OR

Lilium maritimum, CA

Lilium parryi, CA

Prosartes parvifolia, CA, OR

Scoliopus bigelovii, CA, OR

LIMNANTHACEAE

Limnanthes alba, CA, OR

Limnanthes bakeri, CA

Limnanthes montana, CA

Limnanthes vinculans, CA

LINACEAE

Hesperolinon adenophyllum, CA

Hesperolinon bicarpellatum, CA

Hesperolinon breweri, CA

Hesperolinon californicum, CA

Hesperolinon clevelandii, CA

Hesperolinon congestum, CA

Hesperolinon didymocarpum, CA

Hesperolinon disjunctum, CA

Hesperolinon drymarioides, CA

Hesperolinon spergulinum, CA

Hesperolinon tehamense, CA

LOASACEAE

Mentzelia crocea, CA

Mentzelia gracilenta, CA

Mentzelia lindleyi, CA

Mentzelia micrantha, CA

Mentzelia pectinata, CA

LYTHRACEAE

Rotala ramosior, CA

MALVACEAE

Eremalche parryi, CA

Fremontodendron decumbens, CA

Fremontodendron mexicanum, CA, Baja

Malacothamnus abbottii, CA

Malacothamnus aboriginum, CA

Malacothamnus clementinus, CA

Malacothamnus davidsonii, CA

Malacothamnus hallii, CA

Malacothamnus foliosus, Baja

Malacothamnus jonesii, CA

Malacothamnus marrubioides, CA

Malacothamnus palmeri, CA

Malacothamnus paniculatus, Baja

Malva assurgentiflora, CA

Malva lindsayi, Baja

MALVACEAE Continued

Malva occidentalis, Baja

Malva pacifica, Baja

Sidalcea calycosa, CA

Sidalcea celata, CA

Sidalcea diploscypha, CA

Sidalcea elegans, CA, OR

Sidalcea gigantea, CA

Sidalcea glaucescens, CA

Sidalcea hartwegii, CA

Sidalcea hickmanii, CA, OR

Sidalcea hirsuta, CA

Sidalcea keckii, CA

Sidalcea malviflora, CA, Baja

Sidalcea pedata, CA

Sidalcea ranunculacea, CA

Sidalcea reptans, CA

Sidalcea robusta, CA

Sidalcea setosa, CA, OR

Sidalcea sparsifolia, CA, Baja

Sidalcea stipularis, CA

Sphaeralcea fulva, Baja

Sphaeralcea munroana, CA

Sphaeralcea palmeri, Baja

Sphaeralcea sulphurea, Baja

MELANTHIACEAE

Toxicoscordion exaltatum, CA

Toxicoscordion fontanum, CA

Trillium chloropetalum, CA

Veratrum fimbriatum, CA

MONTIACEAE

Calandrinia breweri, CA

Calyptridium pulchellum, CA

Calyptridium pygmaeum, CA

Calyptridium quadripetalum, CA

Cistanthe maritima, CA, Baja

Claytonia gypsophiloides, CA

Claytonia palustris, CA

Claytonia saxosa, CA, OR

Lewisia cantelovii, CA

Lewisia congdonii, CA

Lewisia disepala, CA

Lewisia kelloggii, CA

Lewisia leeana, CA, OR

Lewisia longipetala, CA

Lewisia oppositifolia, CA, OR

Lewisia serrata, CA

Lewisia stebbinsii, CA

NARTHECIACEAE

Narthecium californicum, CA, OR

NYCTAGINACEAE

Abronia alpina, CA

Abronia maritima, CA, Baja

Mirabilis greenei, CA

OLEACEAE

Fraxinus dipetala, CA, Baja

Fraxinus parryi, CA, Baja

Hesperelaea palmeri, Baja

ONAGRACEAE

Camissonia benitensis, CA

Camissonia integrifolia, CA

Camissonia lacustris, CA

Camissonia sierrae, CA

Camissoniopsis bistorta, CA, Baja

Camissoniopsis cheiranthifolia, CA, OR, Baja

Camissoniopsis confusa, CA

Camissoniopsis guadalupensis, CA, Baja

Camissoniopsis hardhamiae, CA

Camissoniopsis hirtella, CA, Baja

Camissoniopsis ignota, CA, Baja

Camissoniopsis intermedia, CA, Baja

ONAGRACEAE Continued

Camissoniopsis lewisii, CA, Baja

Camissoniopsis luciae, CA

Camissoniopsis micrantha, CA

Camissoniopsis robusta, CA, Baja

Clarkia affinis, CA

Clarkia arcuata, CA

Clarkia australis, CA

Clarkia biloba, CA

Clarkia borealis, CA

Clarkia bottae, CA

Clarkia breweri, CA

Clarkia concinna, CA

Clarkia cylindrica, CA

Clarkia davyi, CA

Clarkia delicata, CA, Baja

Clarkia dudleyana, CA

Clarkia epilobioides, CA, Baja

Clarkia exilis, CA

Clarkia franciscana, CA

Clarkia heterandra, CA, OR

Clarkia imbricata, CA

Clarkia jolonensis, CA

Clarkia lewisii, CA

Clarkia lingulata, CA

Clarkia mildrediae, CA

Clarkia modesta, CA

Clarkia mosquinii, CA

Clarkia prostrata, CA

Clarkia purpurea, CA, OR

Clarkia rostrata, CA

Clarkia rubicunda, CA

Clarkia similis, CA

Clarkia speciosa, CA

Clarkia springvillensis, CA

Clarkia stellata, CA

Clarkia tembloriensis, CA

Clarkia unguiculata, CA

Clarkia virgata, CA

Clarkia williamsonii, CA

Clarkia xantiana, CA

Epilobium cleistogamum, CA

Epilobium howellii, CA

Epilobium nivium, CA

Epilobium septentrionale, CA

Epilobium siskiyouense, CA, OR

Gayophytum eriospermum, CA

Gayophytum oligospermum, CA

Oenothera xylocarpa, CA

Tetrapteron graciliflorum, CA, OR, Baja

OPHIOGLOSSACEAE

Botrychium pinnatum, CA

Ophioglossum californicum, CA

ORCHIDACEAE

Corallorhiza trifida, CA

Malaxis monophyllos, CA

Piperia colemanii, CA

Piperia cooperi, CA

Piperia michaelii, CA

Piperia yadonii, CA

Platanthera yosemitensis, CA

OROBANCHACEAE

Castilleja ambigua, CA

Castilleja brevilobata, CA, OR

Castilleja brevistyla, CA

Castilleja cinerea, CA

Castilleja densiflora, CA

Castilleja fruticosa, Baja

Castilleja gleasoni, CA

Castilleja grisea, CA

Castilleja guadalupensis, Baja

Castilleja hololeuca, CA, Baja

Castilleja lasiorhyncha, CA

OROBANCHACEAE Continued

Castilleja latifolia, CA

Castilleja lemmonii, CA

Castilleja lineariloba, CA

Castilleja mollis, CA

Castilleja montigena, CA

Castilleja ophiocephala, Baja

Castilleja praeterita, CA

Castilleja rubicundula, CA

Castilleja schizotricha, CA, OR

Castilleja subinclusa, CA

Castilleja wightii, CA

Chloropyron molle, CA

Chloropyron palmatum, CA

Cordylanthus involutus, Baja

Cordylanthus nevinii, CA, Baja

Cordylanthus nidularius, CA

Cordylanthus pilosus, CA

Cordylanthus pringlei, CA

Cordylanthus rigidus, CA, Baja

Dicranostegia orcuttiana, CA, Baja

Orobanche bulbosa, CA, Baja

Orobanche valida, CA

Orobanche vallicola, CA

Orthocarpus pachystachyus, CA

Pedicularis contorta, CA

Pedicularis dudleyi, CA

Pedicularis howellii, CA, OR

Pedicularis semibarbata, CA

Triphysaria floribunda, CA

Triphysaria micrantha, CA

Triphysaria versicolor, CA

OXALIDACEAE

Oxalis californica, CA, Baja

Oxalis pilosa, CA, Baja

PAEONIACEAE

Paeonia californica, CA, Baja

PAPAVERACEAE

Dendromecon harfordii, CA

Dendromecon rigida, CA, Baja

Dicentra nevadensis, CA

Dicentra pauciflora, CA

Ehrendorferia chrysantha, CA

Ehrendorferia ochroleuca, CA

Eschscholzia caespitosa, CA, OR

Eschscholzia elegans, Baja

Eschscholzia hypecoides, CA

Eschscholzia palmeri, Baja

Eschscholzia lemmonii, CA

Eschscholzia lobbii, CA

Eschscholzia ramosa, CA, Baja

Eschscholzia rhombipetala, CA

Hesperomecon linearis, CA

Meconella californica, CA

Meconella denticulata, CA

Meconella oregana, CA, OR

Papaver californicum, CA

Papaver heterophyllum, CA, Baja

Romneya coulteri, CA, Baja

Romneya trichocalyx, CA, Baja

PHRYMACEAE

Diplacus stellatus, Baja

Mimulus angustatus, CA

Mimulus bicolor, CA

Mimulus bolanderi, CA, OR

Mimulus clevelandii, CA, Baja

Mimulus congdonii, CA, OR

Mimulus constrictus, CA

Mimulus exiguus, CA

Mimulus filicaulis, CA

Mimulus glaucescens, CA

Mimulus gracilipes, CA

Mimulus inconspicuus, CA

Mimulus johnstonii, CA

PHRYMACEAE Continued

Mimulus laciniatus, CA

Mimulus latidens, CA

Mimulus latifolius, CA

Mimulus leptaleus, CA

Mimulus norrisii, CA

Mimulus nudatus, CA

Mimulus pictus, CA

Mimulus pulchellus, CA

Mimulus purpureus, CA, Baja

Mimulus rattanii, CA

Mimulus torreyi, CA

Mimulus traskiae, CA

Mimulus viscidus, CA

Mimulus whitneyi, CA

PICRODENDRACEAE

Tetracoccus dioicus, CA, Baja

PINACEAE

Abies bracteata, CA

Picea breweriana, CA, OR

Pinus balfouriana, CA, OR

Pinus coulteri, CA, Baja

Pinus muricata, CA, Baja

Pinus quadrifolia, CA, Baja

Pinus radiata, CA, Baja

Pinus torreyana, CA

Pseudotsuga macrocarpa, CA

PLANTAGINACEAE

Antirrhinum cornutum, CA

Antirrhinum kelloggii, CA

Antirrhinum leptaleum, CA

Antirrhinum multiflorum, CA

Antirrhinum nuttallianum, CA, Baja

Antirrhinum ovatum, CA

Antirrhinum subcordatum, CA

Antirrhinum vexillocalyculatum, CA, OR

Antirrhinum virga, CA

Callitriche longipedunculata, CA

Callitriche trochlearis, CA

Collinsia antonina, CA

Collinsia childii, CA

Collinsia concolor, CA, Baja

Collinsia corymbosa, CA

Collinsia greenei, CA

Collinsia heterophylla, CA, Baja

Collinsia multicolor, CA

Collinsia tinctoria, CA

Gambelia speciosa, CA, Baja

Keckiella cordifolia, CA, Baja

Keckiella corymbosa, CA

Keckiella lemmonii, CA, OR

Keckiella ternata, CA, Baja

Penstemon caesius, CA

Penstemon californicus, CA, Baja

Penstemon cedrosensis, Baja

Penstemon filiformis, CA

Penstemon heterophyllus, CA

Penstemon labrosus, CA, Baja

Penstemon neotericus, CA

Penstemon newberryi, CA, OR

Penstemon parvulus, CA, OR

Penstemon personatus, CA

Penstemon purpusii, CA

Penstemon spectabilis, CA, Baja

Penstemon tracyi, CA

Plantago erecta, CA, OR, Baja

Veronica copelandii, CA

Veronica cusickii, CA

PLUMBAGINACEAE

Limonium californicum, CA, OR, Baja

POACEAE

Agrostis blasdalei, CA

Agrostis densiflora, CA

Agrostis elliottiana, CA

Agrostis hendersonii, CA, OR

Agrostis hooveri, CA

Alopecurus carolinianus, CA

Aristida divaricata, CA

Bromus grandis, CA

Bromus hallii, CA

Bromus pseudolaevipes, CA

Calamagrostis bolanderi, CA

Calamagrostis breweri, CA

Calamagrostis foliosa, CA

Calamagrostis koelerioides, CA, OR

Calamagrostis muiriana, CA

Calamagrostis ophitidis, CA

Cinna bolanderi, CA

Dissanthelium californicum, CA

Elymus californicus, CA

Elymus pacificus, CA

Elymus sierrae, CA

Elymus stebbinsii, CA

Festuca bajacaliforniana, Baja

Festuca viridula, CA, OR

Hordeum intercedens, CA

Melica torreyana, CA

Muhlenbergia californica, CA

Muhlenbergia jonesii, CA

Muhlenbergia montana, CA

Muhlenbergia utilis, CA

Neostapfia colusana, CA

Orcuttia californica, CA, Baja

Orcuttia inaequalis, CA

Orcuttia pilosa, CA

Orcuttia viscida, CA

Phalaris californica, CA, OR

Pleuropogon californicus, CA

Pleuropogon hooverianus, CA

Poa atropurpurea, CA

Poa bajaensis, Baja

Poa diaboli, CA

Poa douglasii, CA

Poa kelloggii, CA

Poa napensis, CA

Poa pringlei, CA, OR

Poa rhizomata, CA, OR

Poa sierrae, CA

Poa stebbinsii, CA

Poa tenerrima, CA

Puccinellia howellii, CA

Spartina foliosa, CA, Baja

Stipa bracteata, Baja

Stipa cernua, CA, Baja

Stipa coronata, CA, Baja

Stipa diegoensis, CA, Baja

Stipa kingii, CA

Stipa latiglumis, CA

Stipa lepida, CA, Baja

Stipa pulchra, CA

Stipa stillmanii, CA

Tuctoria mucronata, CA

POLEMONIACEAE

Allophyllum divaricatum, CA

Allophyllum glutinosum, CA

Collomia diversifolia, CA

Collomia rawsoniana, CA

Collomia tracyi, CA

Eriastrum abramsii, CA

Eriastrum brandegeeae, CA

Eriastrum filifolium, CA, Baja

Eriastrum hooveri, CA

Eriastrum luteum, CA

Eriastrum virgatum, CA

POLEMONIACEAE Continued

Gilia angelensis, CA, Baja

Gilia austro-occidentalis, CA

Gilia interior, CA

Gilia millefoliata, CA, OR

Gilia nevinii, CA, Baja

Gilia tenuiflora, CA

Gilia yorkii, CA

Ipomopsis guttata, Baja

Leptosiphon acicularis, CA

Leptosiphon ambiguus, CA

Leptosiphon androsaceus, CA, Baja

Leptosiphon croceus, CA

Leptosiphon filipes, CA

Leptosiphon grandiflorus, CA

Leptosiphon jamauensis, Baja

Leptosiphon jepsonii, CA

Leptosiphon latisectus, CA

Leptosiphon laxus, Baja

Leptosiphon montanus, CA

Leptosiphon nudatus, CA

Leptosiphon oblanceolatus, CA

Leptosiphon parviflorus, CA, Baja

Leptosiphon pygmaeus, CA

Leptosiphon rattanii, CA

Leptosiphon rosaceus, CA

Leptosiphon serrulatus, CA

Linanthus bellus, CA, Baja

Linanthus californicus, CA

Linanthus concinnus, CA

Linanthus dianthiflorus, CA, Baja

Linanthus jaegeri, CA

Linanthus killipii, CA

Linanthus veatchii, Baja

Navarretia cotulifolia, CA

Navarretia eriocephala, CA

Navarretia gowenii, CA

Navarretia hamata, CA, Baja

Navarretia heterodoxa, CA

Navarretia jepsonii, CA

Navarretia mellita, CA

Navarretia mitracarpa, CA

Navarretia myersii, CA

Navarretia nigelliformis, CA

Navarretia ojaiensis, CA

Navarretia peninsularis, CA

Navarretia prolifera, CA

Navarretia prostrata, CA

Navarretia pubescens, CA, OR

Navarretia rosulata, CA

Navarretia setiloba, CA

Navarretia viscidula, CA

Phlox adsurgens, CA, OR

Phlox dispersa, CA

Phlox dolichantha, CA

Phlox hirsuta, CA

Polemonium eximium, CA

Saltugilia caruifolia, CA, Baja

Saltugilia splendens, CA

POLYGALACEAE

Polygala californica, CA, OR

POLYGONACEAE

Acanthoscyphus parishii, CA

Aconogonon davisiae, CA, OR

Aristocapsa insignis, CA

Chorizanthe angustifolia, CA

Chorizanthe biloba, CA

Chorizanthe blakleyi, CA

Chorizanthe breweri, CA

Chorizanthe clevelandii, CA

Chorizanthe cuspidata, CA

Chorizanthe diffusa, CA

Chorizanthe douglasii, CA

POLYGONACEAE Continued

Chorizanthe fimbriata, CA, Baja

Chorizanthe howellii, CA

Chorizanthe inequalis, Baja

Chorizanthe leptotheca, CA

Chorizanthe membranacea, CA, OR

Chorizanthe obovata, CA

Chorizanthe orcuttiana, CA

Chorizanthe palmeri, CA

Chorizanthe polygonoides, CA, Baja

Chorizanthe procumbens, CA, Baja

Chorizanthe pungens, CA

Chorizanthe rectispina, CA

Chorizanthe robusta, CA

Chorizanthe staticoides, CA

Chorizanthe stellulata, CA

Chorizanthe turbinata, Baja

Chorizanthe uniaristata, CA

Chorizanthe valida, CA

Chorizanthe ventricosa, CA

Chorizanthe wheeleri, CA

Dodecahema leptoceras, CA

Eriogonum alpinum, CA

Eriogonum apricum, CA

Eriogonum arborescens, CA

Eriogonum argillosum, CA

Eriogonum breedlovei, CA

Eriogonum butterworthianum, CA

Eriogonum callistum, CA

Eriogonum cedrorum, CA

Eriogonum cinereum, CA

Eriogonum cithariforme, CA

Eriogonum congdonii, CA

Eriogonum covilleanum, CA

Eriogonum crocatum, CA

Eriogonum dasyanthemum, CA

Eriogonum diclinum, CA, OR

Eriogonum eastwoodianum, CA

Eriogonum elegans, CA

Eriogonum elongatum, CA, Baja

Eriogonum evanidum, CA, Baja

Eriogonum fastigiatum, Baja

Eriogonum giganteum, CA

Eriogonum gossypinum, CA

Eriogonum gracillimum, CA

Eriogonum hastatum, Baja

Eriogonum hirtellum, CA

Eriogonum hirtiflorum, CA

Eriogonum inerme, CA

Eriogonum kelloggii, CA

Eriogonum libertini, CA

Eriogonum luteolum, CA, OR

Eriogonum molestum, CA

Eriogonum molle, Baja

Eriogonum nervulosum, CA

Eriogonum nortonii, CA

Eriogonum ordii, CA

Eriogonum pendulum, CA, OR

Eriogonum polypodum, CA

Eriogonum prattenianum, CA

Eriogonum roseum, CA, OR, Baja

Eriogonum siskiyouense, CA

Eriogonum spectabile, CA

Eriogonum temblorense, CA

Eriogonum ternatum, CA, OR

Eriogonum tripodum, CA

Eriogonum truncatum, CA

Eriogonum twisselmannii, CA

Eriogonum vestitum, CA

Eriogonum zapatoense, Baja

Hollisteria lanata, CA

Lastarriaea coriacea, CA, Baja

Mucronea californica, CA

Polygonum bidwelliae, CA

POLYGONACEAE Continued

Polygonum bolanderi, CA

Polygonum hickmanii, CA

Polygonum marinense, CA

Sidotheca caryophylloides, CA

Sidotheca emarginata, CA

Systenotheca vortriedei, CA

POLYPODIACEAE

Polypodium californicum, CA, Baja

PORTULACACEAE

Cistanthe guadalupensis, Baja

PRIMULACEAE

Dodecatheon clevelandii, CA, Baja

Dodecatheon subalpinum, CA

PTERIDACEAE

Aspidotis californica, CA

Aspidotis carlotta-halliae, CA

Cheilanthes clevelandii, CA

Cheilanthes cooperae, CA

Cheilanthes newberryi, CA, Baja

Pellaea andromedifolia, CA, OR, Baja

Pentagramma pallida, CA

RANUNCULACEAE

Aquilegia eximia, CA

Clematis lasiantha, CA, Baja

Delphinium antoninum, CA

Delphinium bakeri, CA

Delphinium californicum, CA

Delphinium decorum, CA, OR

Delphinium gracilentum, CA

Delphinium gypsophilum, CA

Delphinium hesperium, CA

Delphinium hutchinsoniae, CA

Delphinium inopinum, CA

Delphinium luteum, CA

Delphinium parryi, CA

Delphinium patens, CA, Baja

Delphinium uliginosum, CA

Delphinium umbraculorum, CA

Delphinium variegatum, CA

Enemion occidentale, CA, OR

Ranunculus austrooreganus, OR

Ranunculus bonariensis, CA

Ranunculus canus, CA

Ranunculus hystriculus, CA

RHAMNACEAE

Adolphia californica, CA, Baja

Ceanothus arboreus, CA, Baja

Ceanothus arcuatus, CA, OR

Ceanothus bolensis, Baja

Ceanothus confusus, CA

Ceanothus crassifolius, CA, Baja

Ceanothus cyaneus, CA, Baja

Ceanothus dentatus, CA

Ceanothus divergens, CA

Ceanothus diversifolius, CA

Ceanothus ferrisiae, CA

Ceanothus foliosus, CA

Ceanothus fresnensis, CA

Ceanothus gloriosus, CA

Ceanothus hearstiorum, CA

Ceanothus impressus, CA

Ceanothus incanus, CA

Ceanothus jepsonii, CA

Ceanothus lemmonii, CA

Ceanothus leucodermis, CA, Baja

Ceanothus maritimus, CA

Ceanothus masonii, CA

Ceanothus megacarpus, CA

Ceanothus oliganthus, CA, Baja

Ceanothus ophiochilus, CA

Ceanothus otayensis, CA, Baja

RHAMNACEAE Continued

Ceanothus palmeri, CA, Baja

Ceanothus papillosus, CA, Baja

Ceanothus parvifolius, CA

Ceanothus pinetorum, CA

Ceanothus pumilus, CA, OR

Ceanothus purpureus, CA

Ceanothus rigidus, CA

Ceanothus roderickii, CA

Ceanothus sonomensis, CA

Ceanothus spinosus, CA, Baja

Ceanothus tomentosus, CA, Baja

Ceanothus verrucosus, CA, Baja

Rhamnus crocea, CA, Baja

Rhamnus pilosa, CA

Rhamnus pirifolia, CA, Baja

ROSACEAE

Acaena pinnatifida, CA

Adenostoma fasciculatum, CA, Baja

Adenostoma sparsifolium, CA, Baja

Cercocarpus minutiflorus, CA, Baja

Cercocarpus traskiae, CA

Chamaebatia australis, CA, Baja

Chamaebatia foliolosa, CA

Drymocallis cuneifolia, CA

Drymocallis hansenii, CA

Drymocallis rhomboidea, CA, OR

Horkelia bolanderi, CA

Horkelia californica, CA

Horkelia clevelandii, CA, Baja

Horkelia congesta, CA, OR

Horkelia cuneata, CA, Baja

Horkelia hendersonii, CA, OR

Horkelia marinensis, CA

Horkelia parryi, CA

Horkelia rydbergii, CA

Horkelia sericata, CA, OR

Horkelia tenuiloba, CA

Horkelia truncata, CA

Horkelia tularensis, CA

Horkelia wilderae, CA

Horkelia yadonii, CA

Horkeliella purpurascens, CA

Ivesia argyrocoma, CA

Ivesia callida, CA

Ivesia campestris, CA

Ivesia longibracteata, CA

Ivesia muirii, CA

Ivesia pickeringii, CA

Ivesia pygmaea, CA

Ivesia santolinoides, CA

Ivesia unguiculata, CA

Lyonothamnus floribundus, CA

Neviusia cliftonii, CA

Potentilla cristae, CA

Potentilla grayi, CA

Potentilla hickmanii, CA

Potentilla luteosericea, Baja

Potentilla multijuga, CA

Potentilla rimicola, CA, Baja

Potentilla uliginosa, CA

Potentilla wheeleri, CA

Rosa bridgesii, CA, OR

Rosa californica, CA, OR, Baja

Rosa minutifolia, CA, Baja

Rosa pinetorum, CA

Rubus glaucifolius, CA, OR

RUBIACEAE

Galium andrewsii, CA, Baja

Galium angulosum, Baja

Galium buxifolium, CA

Galium californicum, CA

Galium catalinense, CA

Galium clementis, CA

RUBIACEAE Continued

Galium cliftonsmithii, CA

Galium coronadoense, Baja

Galium diabolense, Baja

Galium grande, CA

Galium grayanum, CA

Galium hallii, CA

Galium hardhamiae, CA

Galium jepsonii, CA

Galium johnstonii, CA

Galium martirense, Baja

Galium nuttallii, CA

Galium porrigens, CA, OR, Baja

Galium sparsiflorum, CA

Galium wigginsii, Baja

RUSCACEAE

Nolina cismontana, CA

Nolina interrata, CA

RUTACEAE

Cneoridium dumosum, CA, Baja

Ptelea aptera, Baja

Ptelea crenulata, CA, OR

SALICACEAE

Salix breweri, CA

Salix delnortensis, CA, OR

Salix tracyi, CA, OR

SAXIFRAGACEAE

Bolandra californica, CA

Boykinia rotundifolia, CA

Heuchera abramsii, CA

Heuchera brevistaminea, CA

Heuchera caespitosa, CA

Heuchera hirsutissima, CA

Heuchera maxima, CA

Heuchera merriamii, CA, OR

Heuchera parishii, CA

Heuchera pilosissima, CA

Jepsonia heterandra, CA

Jepsonia malvifolia, CA

Jepsonia parryi, CA

Lithophragma bolanderi, CA

Lithophragma cymbalaria, CA

Lithophragma heterophyllum, CA

Lithophragma maximum, CA

Micranthes aprica, CA, OR

Micranthes bryophora, CA

SCROPHULARIACEAE

Scrophularia atrata, CA

Scrophularia villosa, CA

SELAGINELLACEAE

Selaginella asprella, CA, Baja

Selaginella cinerascens, CA, Baja

Selaginella hansenii, CA

SMILACACEAE

Smilax californica, CA, OR

Smilax jamesii, CA

SOLANACEAE

Lycium verrucosum, CA

Solanum palmeri, Baja

Solanum wallacei, CA, Baja

STAPHYLEACEAE

Staphylea bolanderi, CA

STYRACACEAE

Styrax redivivus, CA

TAXACEAE

Torreya californica, CA

TECOPHILAEACEAE

Odontostomum hartwegii, CA

THEMIDACEAE

Bloomeria clevelandii, CA

Bloomeria crocea, CA, Baja

Bloomeria humilis, CA

Brodiaea appendiculata, CA

Brodiaea californica, CA, OR

Brodiaea filifolia, CA

Brodiaea insignis, CA

Brodiaea jolonensis, CA, Baja

Brodiaea kinkiensis, CA

Brodiaea leptandra, CA

Brodiaea matsonii, CA

Brodiaea minor, CA, OR

Brodiaea nana, CA

Brodiaea orcuttii, CA

Brodiaea pallida, CA

Brodiaea rosea, CA

Brodiaea santarosae, CA

Brodiaea sierrae, CA

Brodiaea stellaris, CA

Dichelostemma volubile, CA, OR

Triteleia clementina, CA

Triteleia crocea, CA, OR

Triteleia dudleyi, CA

Triteleia guadalupensis, Baja

Triteleia ixioides, CA, OR

Triteleia laxa, CA, OR

Triteleia lilacina, CA

Triteleia lugens, CA

Triteleia montana, CA

Triteleia peduncularis, CA, OR

THYMELAEACEAE

Dirca occidentalis, CA

URTICACEAE

Hesperocnide tenella, CA, Baja

VERBENACEAE

Verbena californica, CA

Verbena moranii, Baja

Verbena orcuttiana, Baja

VIOLACEAE

Viola pedunculata, CA, Baja

Viola pinetorum, CA

Viola tomentosa, CA

Literature Cited

Abbott, J.C. 2007. *Ischnura gemina*. In IUCN Red List of Threatened Species, Version 2011.1.

Ackerly, D.D. 2003. Community assembly, niche conservatism, and adaptive evolution in changing environments. International Journal of Plant Sciences 164: S165–S184.

———. 2009. Evolution, origin, and ages of lineages in the Californian and Mediterranean floras. Journal of Biogeography 36: 1221–33.

Ackerly, D.D., S.R. Loarie, W.K. Cornwell, S.B. Weiss, H. Hamilton, R. Branciforte, and N.J.B. Kraft. 2010. The geography of climate change: implications for conservation. Diversity and Distributions 16: 476–87.

Ackerly, D.D., D.W. Schwilk, and C.O. Webb. 2006. Niche evolution and adaptive radiation: testing the order of trait divergence. Ecology 87: S50–S61.

Agapow, P.M., O.R.P. Bininda-Edmonds, K.A. Crandall, J.L. Gittleman, G.M. Mace, J.C. Marshall, and A. Purvis. 2004. The impact of species concept on biodiversity studies. Quarterly Review of Biology 79: 161–79.

Airey, S.B. 2010. Conservation easements in private practice. Real Property, Trust and Estate Law Journal 44: 745–822.

Alagona, P.S., and S. Pincetl. 2008. The Coachella Valley multiple species habitat conservation plan: a decade of delays. Environmental Management 41: 1–11.

Alexander, E.A., R.G. Coleman, T. Keeler-Wolf, and S. Harrison. 2006. Serpentine geoecology of western North America. Oxford University Press, Oxford.

Allister, D.E., S.P. Platania, F.W. Schueler, M.E. Baldwin, and D.S. Lee. 1986. Ichthyofaunal patterns on a geographic grid. Chapter 2 (pp. 18–51) in C.H. Hocutt and E.O. Wiley (eds.), The zoogeography of North American freshwater fishes. Wiley, New York.

AmphibiaWeb 2011. www. amphibia.org (accessed Nov. 1, 2011).

Anacker, B.L., and S. Harrison 2012. Climate and the evolution of serpentine endemism in California. Evolutionary Ecology 26: 1011–23.

———. 2013. Historical and ecological controls on phylogenetic diversity in Californian plant communities. American Naturalist (in press).

Anacker, B.L., J. Whittall, E. Goldberg, and S. Harrison. 2011. Origins and consequences of serpentine endemism in the California flora. Evolution 65: 365–76.

Araujo, M.B., D. Nogués-Bravo, J.A.F. Diniz-Filho, A.M. Haywood, P.J. Valdes, and C. Rahbek. 2008. Quaternary climate changes explain diversity among reptiles and amphibians. Ecography 31: 8–15.

Babbitt, B. 2005. Cities in the wilderness: A new vision of land use in America. Island Press, Covelo, CA.

Baldwin, B.G. 1993. Molecular phylogenetics of *Calycadenia* (Compositae) based on ITS sequences of nuclear ribosomal DNA: chromosomal and morphological evolution reexamined. American Journal of Botany 80: 222–38.

———. 2003. A phylogenetic perspective on the origin and evolution of Madiinae. Pp. 193–228 in S. Carlquist, B.G. Baldwin, and G.D. Carr (eds.), Tarweeds & silverswords: evolution of the Madiinae. Missouri Botanical Garden Press, St. Louis.

———. 2005. Origin of the serpentine-endemic herb *Layia discoidea* from the widespread *L. glandulosa* (Compositae). Evolution 59: 2473–79.

———. 2006. Contrasting patterns and processes of evolutionary change in the tarweed-silversword lineage: revisiting Clausen, Keck and Hiesey's findings. Annals of the Missouri Botanical Garden 93: 64–93.

———. 2007. Adaptive radiation of shrubby tarweeds (*Deinandra*) in the California Islands parallels diversification of the Hawaiian silversword alliance (Compositae–Madiinae). American Journal of Botany 94: 237–48.

Baldwin, B.G., D.H. Goldman, D.J. Keil, R. Patterson, T.J. Rosatti, and D.H. Wilken (eds.). 2012. The Jepson manual: vascular plants of California. 2nd ed. University of California Press, Berkeley.

Barbour, M.G., T. Keeler-Wolf, and A. Schoenherr. 2007. Terrestrial vegetation of California. 3rd ed. University of California Press, Berkeley.

Bartel, J.S., J.C. Knight, D.R. Elam. 2001. The Endangered Species Act and rare plant protection in California. In CPNS, Inventory of rare and endangered plants of California, 6th ed. CNPS, Sacramento, CA.

Basinger, J.F., D.R. Greenwood, and T. Sweda. 1994. Early Tertiary vegetation of Arctic Canada. Pp. 175–98 in M.C. Boulter and H.C. Fischer (eds.), Cenozoic plants and climates of the Arctic. Springer Verlag, Berlin.

Bay Area Open Space Council. 2011. The conservation lands network: San Francisco Bay Area Upland Habitat Goals Project report. Berkeley, CA.

Beardsley, P.M., S.E. Schoenig, J.B. Whittall, and R.G. Olmstead. 2004. Patterns of evolution in western North American *Mimulus* (Phrymaceae). American Journal of Botany 91: 474–89.

Beier, P., and R.F. Noss. 1998. Do habitat corridors provide connectivity? Conservation Biology 12: 1241–52.

Bell, C.D., and R.W. Patterson. 2000. Molecular phylogeny and biogeography of *Linanthus* (Polemoneaceae). American Journal of Botany 87: 1857–70.

BirdLife International. 2011. Endemic bird area factsheet: California. www .birdlife.org (accessed Sept. 8, 2011).

Bond, J.E., D.A. Beamer, T. Lamb, and M.C. Hedin. 2006. Combining genetic and geospatial analyses to infer population extinction in mygalomorph spiders endemic to the Los Angeles region. Animal Conservation 9: 145–57.

Bond, J.E., M.C. Hedin, M.G. Ramirez, and B.D. Opell. 2001. Deep molecular divergence in the absence of morphological and ecological change in the Californian coastal dune endemic trapdoor spider *Aptostichus simus*. Molecular Ecology 10: 899–910.

Bond, M.L., D.E. Lee, R.B. Siegel, and J.P. Ward. 2010. Habitat use and selection by California spotted owls in a postfire landscape. Journal of Wildlife Management 73: 1116–24.

Bowen, L., and D. Van Vuren. 1997. Insular endemic plants lack defenses against herbivores. Conservation Biology 11: 1249–54.

Brady, K.U., A.R. Kruckeberg, and H.D. Bradshaw. 2005. Evolutionary ecology of plant adaptation to serpentine soils. Annual Review of Ecology Evolution and Systematics 36: 243–66.

Brown, S.M. 1994. Migrations and evolution: computerized maps from computerized data. Pp. 327–46 in M.C. Boulter and H.C. Fischer (eds.), Cenozoic plants and climates of the Arctic. Springer Verlag, Berlin.

Buchholz, D., and T. Hayes. 2002. Evolutionary patterns of diversity in spadefoot toad metamorphosis (Anura: Pelobatidae). Copeia 1: 180–89.

Burge, D.O., D.M. Erwin, M.B. Islam, J. Kellermann, S.W. Kembel, D.H. Wilken, and P.S. Manos. 2011. Diversification of *Ceanothus* (Rhamnaceae) in the California Floristic Province. International Journal of Plant Sciences 172: 1137–64.

Burns, K.J., M.P. Alexander, D.N. Barhoum, and E.A. Sgaraglia. 2007. Statistical tests of congruence among phylogeographic histories of three avian species in the California Floristic Province. Ornithological Monographs 63: 96–109.

Calflora. 2011. Information on California plants for education, research and conservation. www.calflora.org/ (accessed Nov. 29, 2011).

California Department of Fish and Wildlife (CDFW). 2003. Atlas of the biodiversity of California. California Department of Fish and Wildlife, Sacramento.

California Native Plant Society (CNPS). 2001. Inventory of rare and endangered plants of California. 6th ed. CNPS, Sacramento, CA.

Calsbeek, R., J.N. Thompson, and J.E. Richardson. 2003. Patterns of molecular evolution and diversification in a biodiversity hotspot: the California Floristic Province. Molecular Ecology 12: 1021–29.

Cayan, D.R., E.P. Mauer, M.D. Dettinger, M. Tyree, and K. Hayhoe. 2008. Climate change scenarios for the California region. Climatic Change 87: S21–S42.

Chabot, B.F., and W.D. Billings 1972. Origins and ecology of the Sierran alpine flora and vegetation. Ecological Monographs 47: 163–98.

Chatzimanolis, S., and M. S. Caterino 2007. Toward a better understanding of the "Transverse Range break": lineage diversification in Southern California. Evolution 61: 2127–41.

Church, R. L., D. M. Stoms, and F. W. Davis. 1996. Reserve selection as a maximal covering location problem. Biological Conservation 76: 105–12.

Cicero, C. 1996. Sibling species of titmice in the *Parus inornatus* complex (Aves: Paridae). University of California Publication in Zoology 128: 1–217.

Cincotta, R. P., J. Wisnewski, and R. Engelman. 2000. Human population in the biodiversity hotspots. Nature 6781: 990–92.

Clausen, J. 1951. Stages in the evolution of plant species. Cornell University Press, Ithaca, NY.

Comrack, L., B. Bolster, J. Gustafson, D. Steele, and E. Burkett. 2008. Species of Special Concern: a brief description of an important California Department of Fish and Game designation. California Department of Fish and Wildlife, Wildlife Branch, Nongame Wildlife Program Report 2008–03, Sacramento, CA.

Conservation International. 2011. Biodiversity hotspots. California Floristic Province. www.biodiversityhotspots.org/xp/Hotspots/california_floristic / (accessed Nov. 1, 2011).

Cowling, R. M., P. W. Rundel, B. B. Lamont, M. K. Arroyo, and M. Arianotsou. 1996. Plant diversity in mediterranean-climate regions. Trends in Ecology and Evolution 11: 362–66.

Crespi, B. J., and C. P. Sandoval. 2000. Phylogenetic evidence for the evolution of ecological specialization in *Temema* walking-sticks. Journal of Evolutionary Biology 13: 249–62.

Dallman, P. R. 1998. Plant life in the world's mediterranean climates: California, Chile, South Africa, Australia, and the Mediterranean Basin. University of California Press, Berkeley.

Damschen, E. I., S. Harrison, and J. B. Grace. 2010. Climate change effects on an endemic-rich edaphic flora: resurveying Robert H. Whittaker's Siskiyou sites (Oregon, USA). Ecology 91: 3609–19.

Davis, E. B., M. S. Koo, C. Conroy, J. L. Patton, and C. Moritz. 2008. The California Hotspots Project: identifying regions of rapid diversification in mammals. Molecular Ecology 17: 120–38.

Davis, F. W., C. Costello, and D. Stoms. 2006. Efficient conservation in a utility-maximization framework. Ecology and Society 11: 33–50.

Davis, F. W., P. A. Stine, and D. M. Stoms. 1994. Remote sensing and GIS applied to phytogeographic analysis and conservation planning in southwestern California. Journal of Vegetation Science 5: 743–56.

Davis, F. W., P. A. Stine, D. M. Stoms, M. I. Borchert, and A. D. Hollander. 1995. GAP analysis of the actual vegetation of California 1. The southwestern region. Madroño 42: 40–78.

Davis, F. W., D. M. Stoms, A. D. Hollander, K. A. Thomas, P. A. Stine, D. Odion, M. I. Borchert, J. H. Thorne, M. V. Gray, R. E. Walker, K. Warner, and J. Graae. 1998. The California GAP Analysis Project—final report. University of California, Santa Barbara.

Davis, F. W., D. Stoms, J. Scepan, J. Estes, and M. Scott. 1991. An information systems approach to the preservation of biological diversity. International Journal of Geographic Information Systems 4: 55–78.

Davis, M. B., and R. Shaw. 2001. Range shifts and adaptive responses to Quaternary climate change. Science 292: 673–79.

Dawson, M. N. 2001. Phylogeography in coastal marine animals: a solution from California? Journal of Biogeography 28: 723–36.

Delaney, K. S., and R. K. Wayne. 2005. Adaptive units for conservation: population distinction and historic extinctions in the island scrub-jay. Conservation Biology 19: 523–33.

Edwards, S. W. 2004. Paleobotany of California. Four Seasons 12: 1–75.

Ehrlich, P. R., and P. H. Raven. 1964. Butterflies and plants: a study in coevolution. Evolution 18: 586–608.

Ellis, B., D. C. Daly, L. J. Hickey, K. R. Johnson, J. D. Mitchell, P. Wilf, and S. L. Wing. 2009. Manual of leaf architecture. Cornell University Press, Ithaca, NY.

Elmendorf, S. C., and K. A. Moore. 2007. Plant competition varies with community composition in an edaphically complex landscape. Ecology 88: 2640–50.

Emmel, J. F., T. C. Emmel, and S. O. Matoon. 1998. A checklist of the butterflies and skippers of California. Chapter 72 (pp. 825–36) in T. C. Emmel (ed.), Systematics of western North American butterflies. Mariposa Press, Gainesville, FL.

Engler, R., A. Guisan, and L. Rechsteiner. 2004. An improved approach for predicting the distribution of rare and endangered species from occurrence and pseudo-absence data. Journal of Applied Ecology 41: 263–74.

Ernest, H. B., W. M. Boyce, V. C. Bleich, B. May, S. J. Stiver, and S. G. Torres. 2003. Genetic structure of mountain lion (*Puma concolor*) populations in California. Conservation Genetics 4: 353–66.

Erwin, D. M., and H. E. Schorn. 2000. Revision of *Lyonothamnus* A. Gray (Rosaceae) from the Neogene of western North America. International Journal of Plant Sciences 161: 179–93.

Feldman, C. R., and R. F. Hoyer. 2010. A new species of snake in the genus *Contia* (Squamata: Colubridae) from California and Oregon. Copeia 254–67.

Fischer, D. T., C. J. Still, and A. P. Williams. 2009. Significance of summer fog and overcast for drought stress and ecological functioning of coastal California endemic plant species. Journal of Biogeography 36: 783–89.

Fisher, T. W., and R. E. Orth. 1983. The marsh flies of California (Diptera: Sciomyszidae). Bulletin of the California Insect Survey 24: 1–117.

Fitzpatrick, B. M., J. R. Johnson, D. K. Kump, J. J. Smith, S. R. Voss and H. B. Shaffer. 2010. Rapid spread of invasive genes into a threatened native species. Proceedings of the National Academy of Sciences (USA) 107: 3606–10.

Floyd, C. H., D. H. Van Vuren, K. R. Crooks, K. L. Jones, D. K. Garcelon, N. M. Belfiore, J. W. Dragoo, and B. May. 2011. Genetic differentiation of island spotted skunks, *Spilogale gracilis amphiala*. Journal of Mammalogy 92: 148–58.

Fordyce, J., M. Forister, C. C. Nice, J. M. Burns, and A. M. Shapiro. 2008. Patterns of genetic variation between the checkered skippers *Pyrgus communis* and *Pyrgus albescens* (Lepidoperea: Hesperiidae). Annals of the Entomological Society of America 101: 794–800.

Forister M.L. 2005. Independent inheritance of preference and performance in hybrids between host races of *Mitoura* butterflies (Lepidoptera, Lycaenidae). Evolution 59: 1149–55.

Fox, J., and A. Nino-Murcia. 2005. Status of species conservation banking in the United States. Conservation Biology 19: 996–1007.

Franco, G., D. Cayan, A. Luers, M. Hanemass, and B. Croes. 2008. Linking climate change science with policy in California. Climatic Change 87: S7–S20.

Furches, S., L. Wallace, and K. Helenurm. 2009. High genetic divergence characterizes populations of the endemic plant Lithophragma maximum (Saxifragaceae) on San Clemente Island. Conservation Genetics 10: 115–26.

Gall, L.F. 1985. Santa Catalina Island's endemic Lepidoptera. II. The Avalon hairstreak, *Strymon avalona,* and its interaction with the recently introduced gray hairstreak, *Strymon melinus* (Lycaenidae). Pp. 95–104 in A.S. Menke and D.R. Miller (eds.), Entomology of the California Channel Islands: proceedings of the first symposium. Santa Barbara Museum of Natural History, Santa Barbara, CA.

Gervais, B.R., and A.M. Shapiro. 1999. Distribution of edaphic-endemic butterflies in the Sierra Nevada of California. Global Ecology and Biogeography 8: 151–62.

Geyer, R., J.P. Lindner, D.M. Stoms, F.W. Davis, and B. Wittstock. 2010. Coupling GIS and LCA for biodiversity assessments of land use. International Journal of Life Cycle Assessment 15: 692–703.

Goldblatt, P., and J.C. Manning. 2002. Plant diversity of the Cape region of southern Africa. Annals of the Missouri Botanic Garden 89: 281–302.

Goldingay, R.L., P.A. Kelly, and D.F. Williams. 1997. The kangaroo rats of California: endemism and conservation of keystone species. Pacific Conservation Biology 3: 47–60.

Gompert, Z., J.A. Fordyce, M.L. Forister, A.M. Shapiro, and C.C. Nice. 2006. Homoploid hybrid speciation in an extreme habitat. Science 314: 1923–24.

Goodwillie, C. 1999. Multiple origins of self-compatibility in Linanthus section Leptosiphon (Polemoniaceae): phylogenetic evidence from internal-transcribed-spacer sequence data. Evolution 53: 1387–95.

Gottlieb, L.D. 2003. Rethinking classic examples of recent speciation in plants. New Phytologist 161: 71–82.

Grant, V.L. 1981. Plant speciation. 2nd ed. Columbia University Press, New York.

Grigarick, A.A., and L.A. Stange. 1968. The pollen-collecting bees of the Anthidiini of California (Hymenoptera: Megachilidae). Bulletin of the California Insect Survey 9: 1–113.

Hally, M.C., C.J. Basten, and J.H. Willis. 2006. Pleiotropic quantitative trait loci contribute to population divergence in traits associated with life-history variation in *Mimulus guttatus*. Genetics 172: 1829–44.

Hall, M.C., D.B. Lowry, and J.H. Willis. 2010. Is local adaptation in *Mimulus guttatus* caused by trade-offs at individual loci? Molecular Ecology 19: 2739–53.

Hannah, L., G. Midgley, G. Hughes, and B. Bomhard. 2005. The view from the Cape: extinction risk, protected areas, and climate change. BioScience 55: 231–42.

Harden, D. R. 2004. California geology, 2nd ed. Prentice Hall, New York.

Hardig, T. M., P. S. Soltis, and D. E. Soltis. 2000. Diversification of the North American shrub genus *Ceanothus* (Rhamnaceae): conflicting phylogenies from nuclear ribosomal DNA and chloroplast DNA. *American Journal of Botany* 87: 108–23.

Harding, E. K., et al. 2001. The scientific foundations of habitat conservation plans: a quantitative assessment. Conservation Biology 15: 488–500.

Harrison, S., E. I. Damschen, and J. B. Grace. 2010. Ecological contingency in the effects of climate change on forest herbs. Proceedings of the National Academy of Sciences (USA) 107: 19362–67.

Harrison, S., and J. B. Grace. 2007. Biogeographic affinity helps explain the productivity-richness relationship at regional and local scales. American Naturalist 170: S5–S15.

Harrison, S., and N. Rajakaruna. 2011. Serpentine: evolution and ecology of a model system. University of California Press, Berkeley.

Harrison, S., H. D. Safford, J. B. Grace, J. H. Viers, and K. F. Davies. 2006. Regional and local species richness in an insular environment: serpentine plants in California. Ecological Monographs 76: 41–56.

Hawkins, B. A., and E. Porter. 2003. Does herbivore diversity depend on plant diversity? The case of California butterflies. American Naturalist 161: 40–49.

Hayhoe, K., D. Cayan, C. B. Field, P. C. Frumhoff, E. P. Mauer, N. L. Miller, S. C. Moser, S. H. Schneider, K. N. Cahill, E. E. Cleland, L. Dale, R. Drapek, R. M. Hanemann, L. S. Kalkstein, J. Lenihan, C. K. Lunch, R. P. Neilson, S. C. Sheridan, and J. H. Verville. 2004. Emission pathways, climate change, and impacts on California. Proceedings of the National Academy of Sciences (USA) 101: 12422–27.

Helenurm, K., and S. S. Hall. 2005. Dissimilar patterns of genetic variation in two insular endemic plants sharing species characteristics, distribution, habitat, and ecological history. Conservation Genetics 6: 341–53.

Helm, B. P. 1998. Biogeography of eight large branchiopods endemic to Northern California. Pp. 124–39 in C. W. Witham, E. T. Bauder, D. Belk, W. R. Ferren Jr., and R. Ornduff (eds.), Ecology, conservation, and management of vernal pool ecosystems: proceedings from a 1996 conference. California Native Plant Society, Sacramento.

Herbert, T. D., J. D. Schuffert, A. Andreasen, L. Heusser, M. Lyle, A. Mix, A. C. Ravelo, L. D. Stott, and J. C. Herguera. 2001. Collapse of the California Current during glacial maxima linked to climate change on land. Science 293: 71–76.

Hickman, J. C. 1993. The Jepson manual: higher plants of California. University of California Press, Berkeley.

Hilty, J. A., and A. M. Merenlender. 2004. Use of riparian corridors and vineyards by mammalian predators in Northern California. Conservation Biology 18: 126–35.

Huber, P. R., S. E. Greco, and J. H. Thorne. 2010. Spatial scale effects on conservation network design: tradeoffs and omissions in regional versus local-scale planning. Landscape Ecology 25: 683–95.

Huckett, H. C. 1971. The Anthomyiidae of California, exclusive of the subfamily Scatophaginae (Diptera). Bulletin of the California Insect Survey 12: 1–121.

Hurd, P. D., and C. D. Michener. 1955. The megachiline bees of California (Hymenoptera: Megachilidae). Bulletin of the California Insect Survey 3: 1–248.

International Union for the Conservation of Nature (IUCN). 2011. IUCN Red List of Threatened Species. Version 2011.1. www.iucnredlist.org/.

Jennings, M. R., and M. P. Hayes. 1994. Amphibian and reptile species of special concern in California. Report to the California Department of Fish and Wildlife.

Jockusch, E. L., I. Martinez-Solano, R. W. Hansen, and D. B. Wake. 2012. Morphological and molecular diversification of slender salamanders (Caudata: Plethodontidae: *Batrachoseps*) in the southern Sierra Nevada of California with descriptions of two new species. Zootaxa 3190: 1–30.

Kartesz, J. T., and C. A. Meacham. 1999. Synthesis of the North American flora. www.bonap.org/synth.html.

Kay, K. M., K. L. Ward, L. R. Watt, and D. W. Schemske. 2011. Plant speciation. Chapter 4 (pp. 71–97) in S. Harrison and N. Rajakaruna (eds.), Serpentine: the evolution and ecology of a model system. University of California Press, Berkeley.

Keeler-Wolf, T., D. R. Elam, K. Lewis, and S. A. Flint. 1998. California vernal pool assessment: preliminary report. California Department of Fish and Wildlife, Sacramento.

Kelly, A. E., and M. L. Goulden. 2008. Rapid shifts in plant distribution with recent climate change. Proceedings of the National Academy of Sciences (USA) 105: 11823–26.

Kiesecker, J. M., T. Comendant, T. Grandmason, E. Gray, C. Hall, R. Hilsenbeck, P. Kareiva, L. Lozier, P. Naehu, A. Rissman, M. R. Shaw, and M. Zankel. 2007. Conservation easements in context: a quantitative analysis of their use by the Nature Conservancy. Frontiers in Ecology and the Environment 5: 125–30.

Kimsey, L. S. 1996. Status of terrestrial insects. Chapter 26 (pp. 735–42) in Sierra Nevada Ecosystem Project: final report to Congress. Volume II: Assessments and scientific basis for management options. University of California, Davis, Centers for Water and Wildlands Resources.

King, J. L., M. A. Simovich, and R. C. Brusca 1996. Species richness, endemism, and ecology of crustacean assemblages in Northern Californian vernal pools. Hydrobiologia 328: 85–118.

Klausemeyer, K. R., and M. R. Shaw. 2009. Climate change, habitat loss, protected areas and the climate adaptation potential of species in mediterranean ecosystems worldwide. PloS One 4 (7): e6392.

Klein, C. J., C. Steinback, A. J. Scholz, and H. P. Possingham. 2008. Effectiveness of marine reserve networks in representing biodiversity and minimizing impact to fishermen: a comparison of two approaches used in California. Conservation Letters 1: 44–51.

Knowles, L.L., and D. Otte. 2000. Phylogenetic analysis of montane grasshoppers from western North America (Genus *Melanoplus,* Acrididae: Melanoplinae). Annals of the Entomological Society of America 93: 421–31.

Kraft, N.J.B., B.G. Baldwin, and D.D. Ackerly. 2010. Range size, taxon age, and hotspots of neoendemism in the California flora. Diversity and Distributions 16: 403–13.

Kruckeberg, A.R. 1954. The ecology of serpentine soils: a symposium. III. Plant species in relation to serpentine soils. Ecology 35: 267–74.

———. 1984. California's serpentines: flora, vegetation, geology, soils, and management problems. University of California Press, Berkeley.

———. 2006. Introduction to California soils and plants: serpentine, vernal pools, and other geobotanical wonders. University of California Press, Berkeley.

Kruckeberg, A.R., and D. Rabinowitz. 1985. Biological aspects of endemism in higher plants. Annual Review of Ecology and Systematics 16: 447–79.

Kuchta, S.R. 2007. Contact zones and species limits: hybridization between lineages of the California newt, *Taricha torosa,* in the southern Sierra Nevada. Herpetologica 63: 332–50.

Kuchta, S.R., and A.M. Tan. 2006. Lineage diversification in an evolving landscape: phylogeography of the California newt, *Taricha torosa.* Biological Journal of the Linnean Society 89: 213–39.

Lamoreux, J.F., J.C. Morrison, T.H. Ricketts, D.M. Olson, E. Dinerstein, M.W. McKnight, and H.H. Shugart. 2006. Global tests of biodiversity concordance and the importance of endemism. Nature 440: 212–14.

Lancaster, L.T., and K.M. Kay. 2013. Origin and diversification of the California flora. Evolution (in press).

Lapointe, F.-J., and L.J. Rissler. 2005. Congruence, consensus, and the comparative phylogeography of codistributed species in California. American Naturalist 166: 290–99.

Lavergne, S., J.D. Thompson, E. Garnier, and M. Debusche. 2004. The biology and ecology of narrow endemic and widespread plants: a comparative study of trait variation in 20 congeneric pairs. Oikos 107: 505–18.

Law, J.H., and B.J. Crespi. 2002. The evolution of geographic parthenogenesis in *Timema* walking-sticks. Molecular Ecology 11: 1471–89.

Leaché, A.D., M.S. Koo, C.L. Spencer, T.J. Papenfuss, R.N. Fisher, and J.A. McGuire. 2009. Quantifying ecological, morphological, and genetic variation to delimit species in the coast horned lizard species complex (*Phrynosoma*). Proceedings of the National Academy of Sciences (USA) 106 (30): 12418–23.

Ledig, F.T., and M.T. Conkle. 1983. Gene diversity and genetic structure in a narrow endemic, Torrey pine (*Pinus torreyana* Parry ex Carr). Evolution 37: 79–85.

Lee, S., C. Parr, Y. Hwang, D. Mindell, and J. Choe. 2003. Phylogeny of magpies (genus *Pica*) inferred from mtDNA data. Molecular Phylogenetics and Evolution 29: 250–57.

Lenihan, J.M., D. Bachelet, R.P. Neilson, and R. Drapek. 2008. Response of vegetation distribution, ecosystem productivity, and fire to climate change scenarios in California. Climatic Change 87: S215–S230.

Levin, D. A. 2004. The ecological transition in speciation. New Phytologist 161: 91–96.

Light, T., D. C. Erman, C. Myrick, and J. Clarke. 1995. Decline of the Shasta crayfish (*Pacifastacus fortis* Faxon) of northeastern California. Conservation Biology 9: 1567–77.

Loarie, S. R., B. E. Carter, K. Hayhoe, S. McMahon, R. Moe, C. A. Knight, and D. D. Ackerly. 2008. Climate change and the future of California's endemic flora. PLoS One 3: 2502.

Mac, M. J., P. A. Opler, C. E. Puckett Haecker, and P. D. Doran. 1988. Status and trends of the nation's biological resources. Vol. 2. U.S. Geological Survey, Reston, VA.

Macey, J. R., J. L. Strasburg, J. A. Brisson, V. T. Vredenburg, M. Jennings, and A. Larson. 2001. Molecular phylogenetics of western North American frogs of the *Rana boylii* species group. Molecular Phylogenetics and Evolution 19: 131–43.

Maldonado, J. E., C. Vila, and R. K. Wayne. 2001. Tripartite genetic subdivisions in the ornate shrew (*Sorex ornatus*). Molecular Ecology 10: 127–47.

Mallet, J. 2001. Species, concepts of. Pp. 427–40 in S. Levin (ed.), Encyclopedia of biodiversity. Academic Press, New York.

Manolis, T. D. 2003. Field guide to dragonflies and damselflies of California. University of California Press, Berkeley.

Manolis, T. D., and K. Biggs. 2008. Additional information on the behavior and morphology of the Exclamation Damsel (*Zonagrion exclamationis*). Argia 20 (2): 7–9.

Martin, J. W., and M. K. Wicksten. 2004. Review and redescription of the freshwater atyid shrimp genus *Syncaris* Holmes, 1900, in California. Journal of Crustacean Biology 24: 447–62.

Mason, H. 1952. A systematic study of the genus *Limnanthes* R. Br. *University of California Publications in Botany* 25: 455–512.

Matocq, M. D. 2002. Phylogeographical structure and regional history of the dusky-footed woodrat, *Neotoma fuscipes*. Molecular Ecology 11: 229–42.

McNeill, C. I., and S. K. Jain. 1983. Genetic differentiation studies and phylogenetic inference in the plant genus *Limnanthes* (section Inflexae). Theoretical and Applied Genetics 66: 257–69.

McPhee, J. 1994. Assembling California. Farrar, Straus & Giroux, New York.

Mead, L. S., D. R. Clayton, R. S. Nauman, D. H. Olso, and M. E. Pfrender. 2005. Newly discovered populations of salamanders from Siskiyou County, California, represent a species distinct from Plethodon stormi. Herpetologica 61: 158–77.

Medail, F., and P. Quezel. 1997. Hot-spots analysis for conservation of plant biodiversity in the Mediterranean Basin. Annals of the Missouri Botanical Garden 84: 112–27.

———. 1999. Biodiversity hotspots in the Mediterranean Basin: setting global conservation priorities. Conservation Biology 13: 1510–13.

Menke, A. S., et al., eds. 1979. The semiaquatic and aquatic Hemiptera of California (Heteroptera: Hemiptera). Bulletin of the California Insect Survey 21: 1–166.

Merenlender, A.M., L. Huntsinger, G. Guthey, and S.K. Fairfax. 2004. Land trusts and conservation easements: who is conserving what for whom? Conservation Biology 18: 65–75.

Meyers, S.C., A. Liston, and R. Meinke. 2010. A molecular phylogeny of *Limnanthes* (Limnanthaceae) and investigation of an anomalous *Limnanthes* population from California, U.S.A. Systematic Botany 35: 552–58.

Millar, C.I. 1996. Tertiary vegetation history. Chapter 5 (pp. 71–122) in Sierra Nevada Ecosystem Project: final report to Congress. Volume II: Assessments and scientific basis for management options. University of California, Davis, Centers for Water and Wildlands Resources.

———. 1999. Evolution and biogeography of *Pinus radiata* with a proposed revision of Quaternary history. New Zealand Journal of Forest Science 29: 335–65.

———. 2012. Geologic, climatic, and vegetation history of California. Pp. 49–67 in B.G. Baldwin, D.H. Goldman, D.J. Keil, R. Patterson, T.J. Rosatti, and D.H. Wilken (eds.), The Jepson manual: vascular plants of California. 2nd ed. University of California Press, Berkeley.

Miller, S.E. 1985. The California Channel Islands—past, present, and future: an entomological perspective. Pp. 3–27 in A.S. Menke and D.R. Miller (eds.), Entomology of the California Channel Islands: proceedings of the first symposium. Santa Barbara Museum of Natural History, Santa Barbara, CA.

Minckley, W.L., D.A. Hendrickson, and C.E. Bond. 1986. Geography of western North American freshwater fishes: description and relationships to intracontinental tectonism. Chapter 14 (pp. 519–614) in C.H. Hocutt and E.O. Wiley (eds.), The zoogeography of North American freshwater fishes. Wiley, New York.

Minnich, R. 2007. Climate, paleoclimate, and paleovegetation. Chapter 2 (pp. 43–70) in M.G. Barbour, T. Keeler-Wolf, and A. Schoenherr (eds.), Terrestrial vegetation of California, 3rd ed. University of California Press, Berkeley.

Moilanen, A., K.A. Wilson, and H.P. Possingham, eds. 2009. Spatial conservation prioritization. Oxford University Press, Oxford.

Mora, C., D.P. Tittensor, S. Adt, A.G.B. Simpson, and B. Worm. 2011. How many species are there on Earth and in the ocean? PLoS Biology 9 (8): e1001127.doi.10.1371/journal.pbio.1001127.

Moritz, C., J.L. Patton, C.J. Conroy, J.L. Parra, G.C. White, and S.R. Beissinger. 2008. Impact of a century of climate change on small-mammal communities in Yosemite National Park, USA. Science 322: 261–64.

Morrison, S.A., and W.M. Boyce. 2009. Conserving connectivity: some lessons from mountain lions in Southern California. Conservation Biology 23: 275–85.

Moyle, P.B. 2002. Inland fishes of California. University of California Press, Berkeley.

Moyle, P.B., J.V.E. Katz, and R.M. Quinones. 2011. Rapid decline of California's native inland fishes: a status assessment. Biological Conservation 144: 2414–23.

Moyle, P.B., J.D. Kiernan, P.L. Crain, and R.M. Quinones. 2011. Projected effects of future climates on freshwater fishes of California. California Energy Commission. Publication CEC-500-2011-XXX.

Moyle, P.B., and J.E.Williams. 1990. Biodiversity loss in the temperate zone: decline of the native fish fauna of California. Conservation Biology 4: 275–84.

Moyle, P.B., and R.M. Yoshiyama. 1995. Protection of aquatic biodiversity in California: a five-tiered approach. Fisheries 19: 6–18.

Myers, N., R.A. Mittermeier, C.G. Mittermeier, G.A.B. Da Fonseca, and J. Kent. 2000. Biodiversity hotspots for conservation priorities. Nature 403: 853–58.

Nice, C.C., and A.M. Shapiro. 2001. Population genetic evidence of restricted gene flow between host races in the butterfly genus *Mitoura* (Lepidoptera: Lycaenidae). Annals of the Entomological Society 94: 257–67.

Noss, R.F., M.A. O'Connell, and D.D. Murphy. 1997. The science of conservation planning: habitat conservation under the Endangered Species Act. Island Press, Covelo, CA.

Oberbauer, T. 2002. Analysis of vascular plants species diversity of the Pacific Coast islands of Alta and Baja California. Pp. 201–9 in D.R. Browne, K.L. Mitchell, and H.W. Chaney (eds.), Proceedings of the fifth Channel Islands symposium. Santa Barbara Museum of Natural History, Santa Barbara, CA.

O'Brien, B.C., J. Delgadillo-Rodríquez, S.A. Junak, T.A. Oberbauer, J.P. Rebman, H. Riemann, and S.E. Vanderplank. 2013. The rare, endangered and endemic plants of the California Floristic Province portion of Baja California, Mexico. Aliso (in press).

O'Dell, R.E., and N. Rajakaruna. 2011. Intraspecific variation, adaptation, and evolution. Chapter 5 (pp. 97–137) in S. Harrison and N. Rajakaruna (eds.), Serpentine: the evolution and ecology of a model system. University of California Press, Berkeley.

Oliver, J.C., and A.M. Shapiro. 2007. Genetic isolation and cryptic variation within the *Lycaena xanthoides* species group (Lepidoptera: Lycaenidae). Molecular Ecology 16: 4308–20.

Omland, K.E., C.L. Tarr, W.I. Boarman, J.M. Marzluff, and R.C. Fleischer. 2000. Cryptic genetic variation and paraphyly in ravens. Proceedings of the Royal Society B 267: 2475–82.

Orme, C.D.L., R.G. Davies, M. Burgess, F. Eigenbrod, N. Pickup, V.A. Olson, A.J. Webster, T.S. Ding, P.C. Rasmussen, R.S. Ridgley, A.J. Stattersfield, P.M. Bennett, T.M. Blackburn, K.J. Gaston, and I.P.F. Owens. 2005. Global hotspots of species richness are not congruent with endemism or threat. Nature 436: 1016–19.

Ornduff, R. 1971. Systematic studies of Limnanthaceae. Madroño 21: 103–11.

Parmesan, C. 1966. Climate and species' range. Nature 382: 765–66.

———. 2006. Ecological and evolutionary responses to recent climate change. Annual Review of Ecology, Evolution and Systematics 37: 637–69.

Patton, J.L., and M.F. Smith. 1990. The evolutionary dynamics of the pocket gopher *Thomomys bottae,* with emphasis on populations in California. UC Publications in Zoology 123: 1–161.

Paulson, D.R. 2007. *Zoniagrion exclamationis*. In IUCN Red List of Threatened Species. www.iucnredlist.org/.

Peinado, M., M.A. Macias, J.L. Aguirre, and J. Delgadillo. 2009. A phytogeographical classification of the North American Pacific Coast based on climate, vegetation, and a floristic analysis of vascular plants. Journal of Botany doi:10.1155/2009/389414.

Pelham, J.P. 2008. A catalogue of the butterflies of the United States and Canada, with a complete bibliography of the descriptive and systematic literature. Journal of Research on the Lepidoptera 40: 1–672.

Philips, S.J., and P.W. Comus. 2000. A natural history of the Sonoran Desert. Arizona–Sonora Desert Museum Press, Tucson, AZ.

Platnick, N.I., and D. Ubick. 2008. A revision of the endemic California spider genus *Titiotus* Simon (Araneae, Tengellidae). American Museum Novitiates 3608. 33 pp.

Powell, J.A. 1985. Faunal affinities of Channel Islands Lepidoptera: a preliminary overview. Pp. 69–94 in A.S. Menke and D.R. Miller (eds.), Entomology of the California Channel Islands: proceedings of the first symposium. Santa Barbara Museum of Natural History, Santa Barbara, CA.

———. 1994. Biogeography of Lepidoptera on the California Channel Islands. Pp. 458–64 in W.L. Halvorson and G.J. Maender (eds.), The fourth Channel Islands symposium: update on the status of resources. Santa Barbara Museum of Natural History, Santa Barbara, CA.

Powell, J.A., and Y.-F. Hsu. 2004. Annotated list of California Microlepidoptera. http://essig.berkeley.edu/.

Powell, J.A., and P.A. Opler. 2009. Moths of western North America. University of California Press, Berkeley.

Pratt, G.F. 1994. Evolution of *Euphilotes* (Lepidoptera: Lycaenidae) by seasonal and host shifts. Biological Journal of the Linnean Society 51: 387–416.

Pratt, G.F., and G.R. Ballmer. 1991. Three biotypes of *Apodemia mormo* in the Mojave Desert. Journal of the Lepidoteran Society 45: 46–57.

———. 1993. Correlations of diapause intensities of *Euphilotes* spp. and *Philotiella speciosa* (Lepidoptera: Lycaenidae) with host bloom period and elevation. Annals of the Entomological Society 86: 265–72.

Price, T. 2008. Speciation in birds. Roberts and Co., Boulder, CO.

Ramirez, M.G. 1995. Natural history of the spider genus *Lutica* (Araneae, Zodariidae). Journal of Arachnology 23: 177–93.

Ramirez, M.G., and B. Chi. 2004. Cryptic speciation, genetic diversity and gene flow in the California turret spider *Atypoides riversi* (Araneae: Antrodaietidae). Biological Journal of the Linnean Society 82: 27–37.

Ravelo, A.C., and M.W. Wara. 2004. The role of the tropical oceans on global climate during a warm period and a major climate transition. Oceanography 17 (3): 32–41.

Raven, P.J. 1967. The floristics of the California Islands. Pp. 57–67 in R.N. Philbrick (ed.), Proceedings of the symposium on the biology of the California Islands, 1965. Santa Barbara Botanic Garden, Santa Barbara, CA.

Raven, P.J., and D.I. Axelrod. 1978. Origin and relationships of the California flora. University of California Press, Berkeley.

Richardson, J.E., R.T. Pennington, T.D. Pennington, and P.M. Hollingsworth. 2001. Rapid diversification of a species-rich genus of Neotropical rain forest trees. Science 293: 2242–45.

Richmond, J.Q., and T.W. Reeder 2002. Evidence for parallel ecological speciation in scincid lizards of the *Eumeces skiltonianus* species group (Squamata: Scincidae). Evolution 56: 1498–1513.

Ricketts, T.H., E. Dinerstein, D. Olson, C. Loucks, and W. Eichbaum. 1999. Terrestrial ecoregions of North America: a conservation assessment. Island Press, Washington, DC.

Riemann, H., and E. Ezcurra. 2007. Endemic regions of the vascular flora of the peninsula of Baja California, Mexico. Journal of Vegetation Science 18: 327–36.

Rissler, L.J., R.J. Hijmans, C.H. Graham, C. Moritz, and D.B. Wake. 2006. Phylogeographic lineages and species comparisons in conservation analyses: a case study of Californian herpetofauna. American Naturalist 167: 655–66.

Rissman, A.R., and A.M. Merenlender. 2008. The conservation contributions of conservation easements: analysis of the San Francisco Bay Area protected lands spatial database. Ecology and Society 13 (1): 40. www.ecologyandsociety.org/vol13/iss1/art40/.

Rissman, A.R., R. Reiner, and A.M. Merenlender. 2007. Monitoring natural resources on rangeland conservation easements. Rangelands 29: 21–26.

Rodriguez-Robles, J.A., D.F. Denardo, and R.E. Staub. 2010. Phylogeography of the California mountain kingsnake, *Lampropeltis zonata* (Colubridae). Molecular Ecology 8: 1923–34.

Rodriguez-Robles, J.A., G.R. Stewart, and T.J. Papenfuss. 2001. Mitochondrial DNA-based phylogeography of North American rubber boas, *Charina bottae* (Serpentes: Boidae). Molecular Phylogenetics and Evolution 18: 227–37.

Rosenzweig, M.L. 1995. Species diversity in space and time. Cambridge University Press, Cambridge.

Safford, H.D., J.H. Viers, and S. Harrison. 2005. Serpentine endemism in the Calfornia flora: a database of serpentine affinity. Madroño 52: 222–57.

Sandel, B., L. Arge, B. Dalsgaard, R.G. Davies, K.J. Gaston, W.J. Sutherland, and J.-C. Svenning. 2011. The influence of Late Quaternary climate-change velocity on species endemism. Science 334: 660–64.

Sandoval, C.P., and B.J. Crespi. 2008. Adaptive evolution of cryptic coloration: the shape of host plants and dorsal stripes in *Timema* walking-sticks. Biological Journal of the Linnean Society 94: 1–5.

Sawyer, J.O., J. Gray, G.J. West, D.A. Thornburgh, R.F. Noss, J.H. Engbeck, B. Marcot, and R. Raymond. 2000. History of redwoods and the redwood forest. Chapter 2 (pp. 7–50) in History, ecology and conservation of the Coast redwoods. Island Press, Washington, DC.

Schemske, D., and P. Bierzychudek. 2001. Perspective: evolution of flower color in the desert annual *Linanthus parryae*: Wright revisited. Evolution 55: 1269–82.

Schmitt, J. 1983. Density-dependent pollinator foraging, flowering phenology, and temporal pollen dispersal patterns in *Linanthus bicolor*. Evolution 37: 1247–57.

Schoenherr, A.A. 1992. A natural history of California. University of California Press, Berkeley.

Schoenherr, A. A., C. R. Feldmeth, and M. J. Emerson. 1999. Natural history of the islands of California. University of California Press, Berkeley.

Schoville, S. D., M. Stuckey, and G. K. Roderick. 2011. Pleistocene origin and population history of a neoendemic alpine butterfly. Molecular Ecology 20: 1233–47.

Scott, J. M., F. W. Davis, B. Csuti, R. Noss, B. Butterfield, C. Groves, H. Anderson, S. Caicco, F. D'Erchia, T. C. Edwards, J. Ulliman, and R. G. Wright. 1993. GAP analysis: a geographic approach to protection of biological diversity. Wildlife Monographs 123: 1–41.

Scott, J. M., F. W. Davis, R. G. McGhie, R. G. Wright, C. Groves, and J. Estes. 2001. Nature reserves: do they capture the full range of America's biological diversity? Ecological Applications 11: 999–1007.

Seabloom, E. W., A. P. Dobson, and D. M. Stoms. 2002. Extinction rates under nonrandom patterns of habitat loss. Proceedings of the National Academy (USA) 99: 11229–34.

Service, R. 2007. Delta blues, California style. Science 317: 442–45.

Shaffer, H. B., G. M. Fellers, S. R. Voss, J. C. Oliver, and G. B. Pauly. 2004. Species boundaries, phylogeography, and conservation genetics of the red-legged frog (*Rana aurora/draytonii*) complex. Molecular Ecology 13: 2667–77.

Shaffer, H. B., G. B. Pauly, J. C. Oliver, and P. C. Trenham. 2004. The molecular phylogenetics of endangerment: cryptic variation and historical phylogeography of the California tiger salamander, *Ambystoma californiense*. Molecular Ecology 13: 3033–49.

Shapiro, A. M. 1996. Status of butterflies. Chapter 27 (pp. 743–57) in Sierra Nevada Ecosystem Project: final report to Congress. Volume II: Assessments and scientific basis for management options. University of California, Davis, Centers for Water and Wildlands Resources.

Shapiro A. M., and M. L. Forister. 2005. Phenological "races" of the *Hesperia colorado* complex (Hesperiidae) on the west slope of the California Sierra Nevada. Journal of the Lepidopterists Society 59: 161–65.

Shapiro, A. M., C. A. Palm, and K. L. Wcislo. 1979. The ecology and biogeography of butterflies of the Trinity Alps and Mount Eddy, Northern California. Journal of Research on the Lepidoptera 18: 69–152.

Shevock, J. R. 1996. Status of rare and endemic plants. Chapter 24 (pages 691–707) in Sierra Nevada Ecosystem Project: final report to Congress. Volume II: Assessments and scientific basis for management options. University of California, Davis, Centers for Water and Wildlands Resources.

Shuford, W. D., and T. Gardali, eds. 2008. California bird species of special concern 2006. Studies of western birds 1. California Department of Fish and Wildlife, Sacramento.

Simovich, M. 1998. Crustacean biodiversity and endemism in California's ephemeral wetlands. Pp. 107–18 in C. W. Witham, E. T. Bauder, D. Belk, W. R. Ferren Jr., and R. Orduff (eds.), Ecology, conservation, and management of vernal pool ecosystems: proceedings from a 1996 conference. California Native Plant Society, Sacramento.

Skinner, M.W., D.P. Tibor, R.L. Bittman, B. Ertter, T.S. Ross, S. Boyd, A.C. Sanders, J.R. Shevock, and D.W. Taylor. 1995. Research needs for conserving California's rare plants. Madroño 42: 211–41.

Smith J.P., and J.O. Sawyer. 1988. Endemic vascular plants of northwestern California and southwestern Oregon. Madroño 35: 54–69.

Smocovitis, V.B. 1997. G. Ledyard Stebbins, Jr., and the evolutionary synthesis (1924–1950). American Journal of Botany 84: 1625–37.

Soltis, D.E., P.S. Soltis, and J.A. Tate. 2004. Advances in the study of polyploidy since Plant Speciation. New Phytologist 161: 173–91.

Spasojevic, M.J., E.I. Damschen, and S. Harrison. 2012. Patterns of seed dispersal syndromes on serpentine soils: examining the roles of habitat patchiness, soil infertility, and correlated functional traits. In Plant ecology and diversity (in press).

Spencer, W.D., P. Beier, K. Penrod, K. Winters, C. Paulman, H. Rustigian-Romsos, J. Strittholt, M. Parisi, and A. Pettler. 2010. California Essential Habitat Connectivity Project: a strategy for conserving a connected California. Prepared for California Department of Transportation, California Department of Fish and Wildlife, and Federal Highways Administration.

Spinks, P.Q., R.C. Thomson, H.B. Shaffer. 2010. Nuclear gene phylogeography reveals the historical legacy of an ancient inland sea on lineages of the western pond turtle, *Emys marmorata* in California. Molecular Ecology 19: 542–56.

Springer, Y.P. 2009. Do extreme environments provide a refuge from pathogens? A phylogenetic test using serpentine flax. American Journal of Botany 96: 2010–21.

Stebbins, G.L. 1950. Variation and evolution in plants. Columbia University Press, New York.

Stebbins, G.L., and G.F. Hrusa. 1995. The North Coast biodiversity arena in Central California: a new scenario for research and teaching processes of evolution. Madroño 42: 269–94.

Stebbins, G.L., and J. Major. 1965. Endemism and speciation in the California flora. Ecological Monographs 35: 1–35.

Stein, B.A., L.S. Kutner, and J.S. Adams. 2000. Precious heritage: the status of biodiversity in the United States. Oxford University Press, Oxford.

Stoms, D.M., M.I. Borchert, M. Moritz, F.W. Davis, and R.L. Church. 1998. A systematic process for selecting research natural areas. Natural Areas Journal 18: 338–49.

Stoms, D.M., J.M. McDonald, and F.W. Davis. 1998. Fuzzy assessment of land suitability for scientific research reserves. Environmental Management 29: 545–58.

Strohecker, H.F., W.W. Middlekauff, and D.C. Rentz. 1968. The grasshoppers of California (Orthoptera: Acridoidea). Bulletin of the California Insect Survey 10: 1–177.

Swenson, N.G., and D.J. Howard. 2005. Clustering of contact zones, hybrid zones, and phylogeographic breaks in North America. American Naturalist 166: 581–91.

Takhtajan, A. 1986. Floristic regions of the world. Translated by T. J. Crovello and A. Cronquist. University of California Press, Berkeley.

Tan, A.-M., and D. B. Wake. 2002. MtDNA phylogeography of the California newt, *Taricha torosa* (Caudata, Salamandridae). Molecular Phylogenetics and Evolution 4: 383–94.

Thorne, J. H., D. Cameron, and J. F. Quinn. 2006. A conservation design for the central coast of California and the evaluation of the mountain lion as an umbrella species. Natural Areas Journal 26: 137–48.

Thorne, J. H., P. R. Huber, E. H. Girvetz, J. Quinn, and M. C. McCoy. 2009. Integration of regional mitigation assessment and conservation planning. Ecology and Society 14: 47.

Thorne, J. H., P. R. Huber, and S. Harrison. 2011. Systematic conservation planning: protecting rarity, representation, and connectivity in regional landscapes. Chapter 15 (pp. 309–27) in S. Harrison and N. Rajakaruna (eds.), Serpentine: the evolution and ecology of a model system. University of California Press, Berkeley.

Thorne J. H., J. A. Kennedy, J. F. Quinn, M. McCoy, T. Keeler-Wolf, and J. Menke. 2004. A vegetation map of Napa County using the Manual of California Vegetation classification and its comparison to other digital vegetation maps. Madroño 51: 343–63.

Thorne, J. H., J. H. Viers, J. Price, and D. M. Stoms. 2009. Spatial patterns of endemic plants in California. Natural Areas Journal 29: 344–66.

Thorne, R. F. 1969. The California islands. Annals of the Missouri Botanical Garden 56: 391–408.

Thorp, R. W., D. S. Horning Jr., and L. L. Dunning. 1983. Bumble bees and cuckoo bumble bees of California. Bulletin of the California Insect Survey 23: 1–79.

Thorp, R. W., and J. M. Leong. 1996. Specialist bee pollinators of showy vernal pool flowers. Pp. 169–76 in C. W. Witham, E. T. Bauder, D. Belk, W. R. Ferren Jr., and R. Ornduff (eds.), Ecology, conservation, and management of vernal pool ecosystems: proceedings from a 1996 conference. California Native Plant Society, Sacramento.

Ubick, D. 1989. The harvestman family Phalangogidae: 1. The new genus *Calicina*, with notes on *Sitalcina* (Opiliones: Laniatores). Proceedings of the California Academy of Sciences 46: 95–136.

———. 1995. On males of Californian *Talanites* (Araneae, Gnaphosidae). Journal of Arachnology 23: 209–11.

Ubick, D., and T. S. Briggs. 2008. The harvestman family Phalangodidae. 6. Revision of the *Sitalcina* complex (Opiliones: Laniatores). Proceedings of the California Academy of Sciences, 4th ser., 59: 1–108.

Underwood, E., K. R. Klausmeyer, R. L. Cox, S. M. Busby, S. A. Morrison, and M. R. Shaw. 2008. Expanding the global network of protected areas to save the imperiled mediterranean biome. Conservation Biology 23: 43–52.

Underwood, E., J. H. Viers, K. R. Klausmeyer, R. L. Cox, and M. R. Shaw. 2009. Threats and biodiversity in the mediterranean biome. Diversity and Distributions 15: 188–97.

Underwood, J. G. 2011. Combining landscape-level conservation planning and biodiversity offset programs: a case study. Environmental Management 47: 121–29.

Vandegast, A.G., A.J. Bohonak, D.B. Weissman, and R.N. Fisher. 2007. Understanding the genetic effects of recent habitat fragmentation in the context of evolutionary history: phylogeography and landscape genetics of a Southern California endemic Jerusalem cricket (Orthoptera: Stenopelmatidae: *Stenopelmatus*). Molecular Ecology 16: 977–92.

Viers, J.H., J.H. Thorne, and J.F. Quinn. 2005. CalJep: A spatial distribution database of Calflora and Jepson plant species. San Francisco Estuary and Watershed Science 4, Issue 1, Article 1. http://repositories.cdlib.org/jmie/sfews/vol4/iss1/art1.

Vredenburg, V.T. 2007. Concordant molecular and phenotypic data delineate new taxonomy and conservation priorities for the endangered mountain yellow-legged frog (Ranidae: *Rana muscosa*). Journal of Zoology 271: 361–74.

Wake, D.B. 1997. Incipient species formation in salamanders of the *Ensatina* complex. Proceedings of the National Academy (USA) 94: 7761–67.

———. 2006. Problems with species: patterns and processes of species formation in salamanders. Annals of the Missouri Botanical Garden 93: 8–23.

Wake, D.B., E.A. Hadly, and D.D. Ackerly. 2009. Biogeography, changing climate, and niche evolution. Proceedings of the National Academy (USA) 106: 19631–36.

Weissman, D.B. 1985. Zoogeography of the Channel Islands Orthoptera. Pp. 61–68 in A.S. Menke and D.R. Miller (eds.), Entomology of the California Channel Islands: proceedings of the first symposium. Santa Barbara Museum of Natural History, Santa Barbara, CA.

Weissman, D.B., A.G. Vandergast, and N. Ueshima. 2008. Jerusalem crickets (Orthoptera: Stenopelmatidae). In J.L. Capinera (ed.), Encyclopedia of entomology. Springer Verlag, Berlin.

Wells, P.V. 1969. The relationship between mode of reproduction and extent of speciation in woody genera of the California chaparral. Evolution 23: 264–67.

Whittaker, R.H. 1960. Vegetation of the Siskiyou Mountains, Oregon and California. Ecological Monographs 30: 279–338.

———. 1961. Vegetation history of the Pacific Coast states and the "central" significance of the Klamath region. Madroño 16: 5–23.

———. 1972. Evolution and measurement of species diversity. Taxon 21: 213–51.

Wiens, J., D. Stralberg, D. Jongsomjit, C.A. Howell, and M.A. Snyder 2009. Niches, models, and climate change: assessing the assumptions and uncertainties. Proceedings of the National Academy of Sciences (USA) 106: 19729–36.

Wiggins, I. 1980. Flora of Baja California. Stanford University Press, Stanford, CA.

Wilson, D.E., and S. Ruff (eds.). 1999. The Smithsonian book of North American mammals. Smithsonian Institution, Washington, DC.

Wolfe, J.A. 1978. A paleobotanical interpretation of Tertiary climates in the Northern Hemisphere. American Scientist 66: 694–703.

Wu, C.A., D.B. Lowry, A.M. Cooley, Y.W. Lee and J.H. Willis. 2007. *Mimulus* is an emerging model system for integration of ecological and genomic studies. Heredity 2007: 1–11.

Zachos, J., M. Pagani, L. Sloan, E. Thomas, and K. Billups. 2001. Trends, rhythms, and aberrations in global climate 65 Ma to present. Science 292: 686–93.

Zachos, J.C., G.R. Dickens, and R.E. Zeebe. 2008. An early Cenozoic perspective on greenhouse warming and carbon-cycle dynamics. Nature 451: 279–83.

Zink, R.M., G.F. Barrowclough, J.L. Atwood, and R.C. Blackwell-Rago. 2000. Genetics, taxonomy, and conservation of the threatened California gnatcatcher. Conservation Biology 14: 1394–1405.

Index